Structural and Thermal Analyses of Deepwater Pipes

Structural and Thermal Analyses of Deepwater Pipes

Humberto Varum · Fulvio Parisi ·
Nicola Tarque · Dora Silveira
Editors

Structural Characterization and Seismic Retrofitting of Adobe Constructions

Experimental and Numerical Developments

Springer

Editors
Humberto Varum
CONSTRUCT-LESE, Faculty of Engineering
University of Porto
Porto, Portugal

Fulvio Parisi
Department of Structures for Engineering and Architecture
University of Naples Federico II
Naples, Italy

Nicola Tarque
GERDIS Research Group, Civil Engineering Division, Department of Engineering
Pontificia Universidad Católica del Perú
San Miguel, Lima, Peru

Dora Silveira
Itecons—Institute for Research and Technological Development for Construction, Energy, Environment and Sustainability
Coimbra, Portugal

ISSN 2194-9832 ISSN 2194-9840 (electronic)
Building Pathology and Rehabilitation
ISBN 978-3-030-74736-7 ISBN 978-3-030-74737-4 (eBook)
https://doi.org/10.1007/978-3-030-74737-4

This Springer imprint is published by the registered company Springer Nature Switzerland AG
The registered company address is: Gewerbestrasse 11, 6330 Cham, Switzerland

Preface

The growing interest for reducing the seismic vulnerability of adobe constructions has stimulated important research programmes at national and international levels. In the last decades, significant theoretical and experimental studies have been carried out to investigate the structural performance of adobe buildings and to develop effective systems for structural rehabilitation and strengthening against natural hazards. Most of research was focussed on adobe structures subjected to earthquake ground motions, leading to a number of interesting solutions either for new structures or for structural retrofit of existing structures. The aforementioned motivated the editors of this book to collect and critically review the most important studies at multiple scales, from material characterization to structural systems assessment, also accounting for actual international concerns, such as economic and environmental sustainability.

The aim of the book is to provide professors, researchers, students, civil engineers, architects and all other interested persons with the most recent developments in (i) material and structural experimental characterization, (ii) structural performance assessment through analysis methods with different levels of sophistication and (iii) seismic retrofitting systems with either traditional or more innovative materials and solutions. Therefore, the book can be used as a textbook in undergraduate and post-graduate courses dealing with vernacular and traditional construction techniques, engineering failure analysis, structural analysis and building rehabilitation, which are usually incorporated in civil engineering and architecture programmes. The target audience of the book also includes engineers, architects, facility managers and public officers involved in conservation and rehabilitation of cultural heritage constructions, historical urban centres and archaeological sites.

The book contains a comprehensive discussion of scientific and technological advances in experimental characterization, numerical simulation and seismic retrofitting of adobe constructions around the world. The outcomes of research activities performed by different institutions in the framework of national and international programmes are presented in a number of chapters authored by worldwide leading experts in the field. This allowed the editors and chapter contributors to delineate possible directions for future investigations aimed at mitigating future losses and preserving cultural heritage for next generations.

Chapter "Behaviour of Adobe Construction in Recent Earthquakes" deals with the seismic performance of adobe constructions in recent strong earthquakes occurred in different regions. The most common types of structural damage are outlined to make a reverse analysis of the seismic behaviour of adobe constructions. This allows the identification of the main structural deficiencies that usually have negative impact on seismic performance and resilience.

Chapter "Mechanical Characterization of Adobe Bricks" is focussed on the experimental characterization of mechanical behaviour and properties of adobe bricks. Code provisions, standards and technical recommendations are reviewed together with testing procedures and data presented in the literature. This provides an overview of the existing knowledge on traditional adobe bricks, allowing the identification of open research issues to investigate and scientific developments to implement in standards and codes.

Chapter "Mechanical Characterization of Adobe Masonry" extends the review of mechanical characterization up to the scale of adobe masonry, addressing experimental studies on compressive, flexural and shear behaviour of the masonry assemblage together with shear behaviour of mortar joints.

Chapter "Quasi-static In-Plane Testing of Adobe Masonry Walls and Structures" focusses on quasi-static testing of adobe masonry walls under in-plane loading. Experimental tests were performed on full-scale adobe specimens, with or without openings, which are representative of traditional adobe constructions in Italy, Peru and Portugal. The discussion of experimental data and observed crack patterns allows book readers to understand the main features of the in-plane behaviour of adobe walls. This provides a basis for more complex dynamic tests as well as numerical simulation and structural retrofitting.

Chapter "Shaking Table Testing of Adobe Masonry Structures" presents a comprehensive review of shaking table tests on adobe walls and buildings. Experimental observations of nonlinear dynamic behaviour and failure modes provide a valuable support to post-earthquake forensic analysis of adobe constructions and effective actions for seismic vulnerability mitigation.

Chapter "Non-destructive (NDT) and Minor-destructive (MDT) Testing Tools to Support the Structural Characterization of Adobe Constructions" deals with non-destructive and minor-destructive testing techniques for multi-scale structural characterization of adobe constructions in terms of geometry, damage and physical-mechanical properties. On-site testing can be complemented with laboratory test procedures discussed in previous chapters. This allows a proper diagnosis of existing adobe constructions, supporting design of repair and strengthening interventions.

Chapter "Seismic Strengthening Techniques for Adobe Construction" is focussed on seismic strengthening of adobe constructions through traditional and innovative systems, which are critically discussed in view of their possible implementation in less developed regions or cultural heritage sites. Major challenges regarding reversibility, durability, cost and technology level associated with strengthening systems are considered. The most relevant studies performed during the last four decades are collected. The review addresses experimental research at multiple structural scales, that is, from masonry wallets and load-bearing walls to entire buildings. Grouting techniques are firstly reviewed for repair of earthquake-damaged adobe structures. Other sections focus on the ability of external strengthening systems to prevent in-plane and/or out-of-plane failure modes of adobe buildings.

Chapter "Numerical Modelling of Adobe Structures" deals with computational strategies for seismic performance assessment of adobe structures. Special emphasis is given to the finite element method, discrete element method and equivalent frame method, which are the most common numerical methods used for masonry constructions. According to the general approach of this book, numerical analysis procedures are discussed at multiple structural scales to investigate the use of each computational strategy for different purposes and applications.

The book ends with Chapter "Research Developments and Needs on Seismic Performance and Strengthening of Adobe Masonry Constructions" where primary findings of theoretical and experimental studies are summarized and challenges for future research are identified. This can foster new research programmes and possible developments in standards and codes, both at national and international levels.

Porto, Portugal — Humberto Varum
Naples, Italy — Fulvio Parisi
San Miguel, Peru — Nicola Tarque
Coimbra, Portugal — Dora Silveira

Contents

Adobe Constructions in the World: A First Overview

Fulvio Parisi, Nicola Tarque, Humberto Varum, and Julio Vargas-Neumann

Abstract Adobe constructions date back thousands of years and are a significant fraction of the world's built heritage, including UNESCO sites. Even though adobe dwellings are still built in some regions, a number of experimental and theoretical studies have delineated the vulnerability levels of these constructions. A significant concentration of adobe constructions in regions with moderate-to-high levels of seismic hazard has been observed as well, resulting in huge human losses during earthquakes. Therefore, the knowledge of the structural performance and seismic reinforcement of adobe constructions provides the basis for their structural vulnerability reduction, allowing disaster risk mitigation and community preparedness. This is also a strategic task to meet sustainability and community resilience goals in multiple countries, where guidelines for both existing and new adobe constructions must be implemented. This chapter provides readers with preliminary information on history, worldwide distribution, construction features and observed performance of adobe constructions for different demand scenarios. These topics are recalled and discussed in more detail in subsequent chapters.

Keywords Adobe constructions · Worldwide distribution · History · Earthquakes · Seismic reinforcement · Structural performance · Damage

---

F. Parisi (✉)
Department of Structures for Engineering and Architecture, University of Naples Federico II, Via Claudio 21, 80125 Naples, Italy
e-mail: fulvio.parisi@unina.it

N. Tarque · J. Vargas-Neumann
GERDIS Research Group, Civil Engineering Division, Department of Engineering, Pontificia Universidad Católica del Perú (PUCP), Av. Universitaria 1801, San Miguel, Lima, Peru
e-mail: sntarque@pucp.edu.pe

J. Vargas-Neumann
e-mail: jvargas@pucp.edu.pe

H. Varum
CONSTRUCT–LESE, Department of Civil Engineering, Faculty of Engineering, University of Porto, R. Dr. Roberto Frias, 4200-465 Porto, Portugal
e-mail: hvarum@fe.up.pt

H. Varum et al. (eds.), *Structural Characterization and Seismic Retrofitting of Adobe Constructions*, Building Pathology and Rehabilitation 20,
https://doi.org/10.1007/978-3-030-74737-4_1

1 Introduction

Since ancient times, earth has been one of the most used building materials all around the world (together with wood and natural stone), particularly in Latin America, Africa, Middle East and Southern Europe. Nowadays, a huge fraction of the world population lives in earthen constructions (Houben and Guillaud 1994; Correia et al. 2016). About 10% of the UNESCO World Heritage sites are totally of partially built with earth construction techniques, including European historic cities such as Cordoba (Spain), Lyon (France), Baku (Azerbaijan) and Safranbolu (Turkey) (WHEAP 2012). In other regions of Africa, Asia and Latin America, earth is still used for new dwellings mainly because of the local availability of materials as well as ease of manufacturing and construction techniques. This allows the involvement of non-specialised workmanship, resulting in low-cost, non-engineered, and unreinforced constructions that may be significantly vulnerable to natural hazards, such as floods and earthquakes. The situation is rather different in developed countries where a revival of interest in earthen materials is mostly driven by modern societal demands for cultural heritage preservation and sustainable development of urban and rural centres (Minke 2006; Berge 2009). Indeed, earthen materials ensure thermo-acoustic comfort to building occupants, architectural compatibility with historical built environments, and low energy consumption, which allow meeting economic and environmental sustainability requirements. These considerations delineate risk management and conservation of adobe constructions as a complex and multi-disciplinary task that requires a comprehensive analysis of information from several scientific and technical perspectives. This motivates the following sections, which move across history, evolution of techniques and structural performance observation of adobe constructions at global scale, as a basis for the other book chapters that deal with detailed topics of interest for structural performance assessment and retrofitting.

2 Adobe Construction Across Multiple Ages and Countries

Earthen construction can be found with at least 9000 years in Mesopotamia and Turkmenistan (Middle East) and 5000 years in Caral (Peru), where the fabrication of earthen elements with formworks was regarded as attractive alternative solution to stones cutting. The earthen techniques include, but are not limited to, *cob or hand molded walls, rammed earth* (also known as *pisé* in French, *tapial* in Spanish, *taipa* in Portuguese and *hangtu* in Chinese), *cob*, *torchis*, *quincha*, *adobe*, and *compressed earth blocks* (CEBs). Among them, adobe is of special interest to many researchers and professionals due to its widespread use and, at the same time, it is the ancestor and counterpart of clay brick masonry (CBM). From an etymological standpoint, the term adobe originates from the Arabic word *aṭ-ṭūb*, which means 'the bricks'. CBM consists of fired clay bricks and mortar joints, whereas traditional adobe is a masonry assemblage composed of sun-dried bricks and earthen mortar joints. Adobe bricks are usually produced with a mixture of soil and water into a prismatic formwork. The bricks are then dried by means of the combined action of air and sunshine. The soil used for the fabrication of adobe bricks is composed of clay, silt and aggregates with different size. The soil-water mixture is often reinforced with either natural or artificial fibres, in order to reduce shrinkage cracking and to increase mechanical strength and durability. Natural fibres are made of, for instance, wool, straw, seaweed, barley, coconut, or sisal. In some instances, the soil-water mixture is stabilised through animal or vegetal stabilizers (e.g. casein, animal glue, oil, latex) or other binders (e.g. lime). The composition of earthen mortar is normally very similar to that of adobe bricks, even though proper fractions of lime and/or sand are added to prevent shrinkage cracking of slightly argillaceous soil. Nonetheless, the mechanical properties of adobe materials and constructions change from one region to another.

As it was mention, one of the first sedentary civilizations was Mesopotamia (which in Greek means the land between rivers), located between the Tigris and the Euphrates. Thousand years ago they used first mud, later hand-moulded adobes, and then adobe made in moulds. In the middle Euphrates area, between 7500 and 5500 BC, moulds of two parallel tables woods appeared that pressed the mud. Then, in the period of El Obied (Iraq, 7500 5700 years ago) the four-sided moulds appeared.

Walls moulded by hand or stacked by layers of small height, with the help of reeds or vegetables, were used in American prehistory. Stonemasonry (with or without mortars) are also prehistoric. Since those times, natural disasters have affected buildings built with earth or stone, if not with great rains and floods, with earthquakes in seismic zones.

In Egypt, adobe mastabas (Egyptian tomb) with a pyramidal trunk shape and rectangular section, began to be built from the first dynastic era of the archaic period (3500 BC) and was the genre of the building that preceded and prepared the construction of the Giza pyramids. Egypt is an intermediate seismic zone. The mastabas did not use seismic reinforcements (Fig. 1).

Fig. 1 Egyptian tomb of Adobe (Mastaba). *Credit* Jairo Martín. http://fueradeclase-blogs.pet.com/

In America, specifically in the central north of Peru and close to the Sechín river —15 km from the Pacific coast-, the first civilizations appeared and the first adobes of ovoid, rectangular, and pyramidal trunk were used (Fig. 2, 3500–1500 BC). In Caral, 5000 years ago, mud-straw adobes (Fig. 3) were used to lighten the constructions, simultaneously with the use of the quincha (framework of wooden posts with mud, cane and fiber crossbars, Fig. 4). Therefore, this is the first real knowledge of seismic resistant construction of earthen buildings.

Fig. 2 Low Sechin. First adobes in America. *Credit* Arqueologíadelperu.com

Fig. 3 Light adobes used in Caral with a large amount of straw, called straw-mud blocks. *Credit* J. Vargas-C. Iwaki

Fig. 4 Timber framework, reeds, and earth, at the top of the Caral pyramid. *Credit* J. Vargas-C. Iwaki

In Caral, techniques that used branches, reeds, fibers (mixed or covered with soil) were also used very early in the building of houses. It is important to highlight the use of natural materials (e.g. plants) as reinforcement systems in earthen walls, conforming orthogonal lattices (vertical and horizontal, or crossed diagonals). These discoveries in Caral (Fig. 5) demonstrate that this civilization was one of the precursors of earthquake-resistant engineering. At the same time, other civilizations in the Middle-East also used natural materials (plants) for strengthening their earthen constructions (Fig. 6).

In Europe, adobe constructions are mostly located in Mediterranean countries such as Portugal, Spain and Italy, and to a lesser extent, in other countries such as France, UK, Germany and Turkey (Correia 2011). In Portugal, adobe has been commonly used until the middle of the twentieth century, particularly in the centre and south coastlines of the country (Fernandes and Portugal 2006; Varum et al. 2011; Tavares et al. 2012). The Aveiro district is an emblematic region where about 40% of the built heritage consists of adobe constructions, most of which having an important cultural, social and architectural value, but at the same time, many of those suffer of a poor state of conservation and structural deficiencies. The majority

Fig. 5 Major Pyramid of Caral, Peru. Masonry of stone, earth, and fibers (3000 BC). *Credit* Christofer Kleihege

Fig. 6 Ziggurat from Uruk (Iraq). The step pyramid of successive layers of adobes and reeds (3200–3000 BC). Mesopotamia plain stone shortage. *Credit* Ancient-Origins

of adobe constructions located in the city of Aveiro are two- or three-storey buildings (some of which influenced by the Art-Nouveau architectural style (Dell'Acqua et al. 2009)) with lime-stabilised adobe bricks. By contrast, rural adobe constructions are typically single-storey buildings. In Spain, earthen constructions are mostly made of adobe and rammed earth (Delgado and Guerrero 2006). Even nowadays, some Spanish buildings are built with adobe walls, such as the two-storey Luis Salazar house in Urueña (Valladolid) that was completed in 2002. Some of the external walls of that building are made of rammed earth, whereas the remaining walls are made of adobe. Adobe constructions are also found in Italy, especially in Sardinia, Abruzzo, Marche and Emilia Romagna (Sanna and Atzeni 2009; Achenza and Sanna 2009). In the alluvial plain of Campidano (Sardinia), approximately 30,000 adobe buildings were constructed until the beginning of the Second World War, but about half of them were demolished. In other Italian regions, some older buildings located in urban areas have load-bearing walls made of stone masonry at the ground floor, stone and/or adobe masonry at intermediate floors, and adobe masonry at upper floors, thus allowing a protection of the walls from rising dump. In rural houses, foundation systems were built with stone masonry whereas load-bearing walls were constructed with rammed earth or adobe masonry walls (Fratini et al. 2011). In Sardinia, traditional adobe constructions were built using adobe bricks reinforced with straw fibres and stabilised with dung or even urine (Parisi et al. 2015). Presently, Sardinian adobe bricks are stabilised with cement. In Turkey, adobe has been widely used in Anatolia since prehistoric times (Tanaçan 2008).

In these two last centuries in South America, most of low-income families use adobe to build non-engineering constructions, particularly in rural areas (Blondet et al. 2003). Adobe buildings are also located in urban areas, where a better quality of construction is usually found. Adobe buildings do not always have a strong foundation system, which is made of medium-size or large blocks connected with mud or coarse mortar. Adobe walls are typically composed of adobe blocks and mud mortar, with mud or mixed mud-gypsum stucco. The roof system is made of wood joists that support corrugated zinc sheets or clay tiles. In Peru, a significant fraction of population lives in adobe buildings (which however reduced from 54 to 36% in last two decades). There are some cities, as in Cusco, were around 80% of constructions are built in adobe. In Lima, Colonial houses were usually built with adobe or clay bricks at the ground floor (also named first floor) and quincha (i.e. wooden frame filled with mud and cane) at the second and third floors.

In Asian countries, such as Afghanistan, Pakistan and Iran, adobe masonry is a traditional construction technology used to build both luxury residential buildings and poor houses. Rural Iranian constructions are typically built using adobe mixed with mud, stone, wood or bricks and concrete blocks (Mousavi et al. 2006). It was estimated that at least 26% of 4 million rural houses were built with adobe and rammed earth walls. The roof of Iranian adobe dwellings consists of vaults and/or domes. Vaulted roofs have a semi-cylindrical shape with a couple of plates or hemispherical caps at their ends (so-called *tharby*). Sometimes, the vault is only part of the roof system and is called *kalil*. Otherwise, the roof may consist of a multiple arch system or wooden beams covered with a mixture of tree branches and mud. In China, adobe bricks were used to build houses since the Shang dynasty (1500–1000 BC), becoming popular during the Han dynasty (206 BC–220 AD) (Wu et al. 2013). Even part of the Great Wall of China was constructed with adobe blocks, whereas fired clay bricks and rammed earth were used elsewhere.

In Africa, a significant regional variability of adobe constructions is observed. For instance, in Morocco, adobe dwellings have different characteristics depending on their relative location with respect to the Atlas Mountains. In the Drâa valley, adobe masonry was typically used for construction of columns and decorative elements both inside the patios and on top of buildings (Baglioni et al. 2010). Adobe is also one of the most used construction techniques in the Huambo province, Angola (Duarte et al. 2017).

3 Structural Features and Performance of Adobe Constructions

The seismic knowledge evolves every day and new geographical areas—that seemed not to have a seismic history—appear. We only know part of the history of the earthquakes or return periods and, therefore, the seismic geographical area is gradually increasing. The earthquakes have shown to be very destructive, making earthen constructions very vulnerable and easy to collapse.

Adobe constructions have a variety of positive and critical features that motivate on one hand their use all around the world, and on the other, the severe damages and losses recorded after natural events (Avrami et al. 2008). There is no doubt that adobe constructions have the following major advantages:

- Low cost of raw materials, transportation and manufacturing.
- Moderate fire resistance and good thermal insulation.
- Good acoustic properties.
- Recyclability and low-energy consumption.

Low cost of transportation and manufacturing are motivated by the local availability of raw materials and ease of construction techniques. All advantages listed above have positive effects in terms of economic and environmental sustainability, allowing thermo-acoustic comfort to building occupants and limited lifecycle cost. Nevertheless, as emphasised above, the apparent ease of construction often has a negative implication on structural safety because of the brittle behaviour of the adobe material. Particularly in less developed regions, inexpert workmanship produces non-engineered constructions that may have high vulnerability to natural hazards because of several deficiencies at material, component and system scales of adobe structures, particularly when it is no introduced a reinforcement system.

From a physical-mechanical standpoint, adobe masonry typically has the following disadvantages:

- Shrinkage during the drying phase, which causes internal and/or surface cracking of adobe bricks and mortar.
- Relatively high mass density.
- Poor strength in compression and brittle behaviour in tension and shear.
- Low resistance to water erosion.
- Fast deterioration under environmental actions.

Moving from material scale to structural scale, adobe constructions have a number of vulnerability sources such as the frequent lack of a foundation system, the severe influence in the safety of the geometric irregularities (either in plan or elevation), and of the absence or ineffectiveness of connections between load-bearing walls, between walls and floors/roof components.

The high mass density may produce significant inertia forces that, only in some cases, can be effectively withstood by adobe structures, due to the combination of poor material properties and construction options. The vulnerability of adobe constructions may further increase in the event of missing or ineffective maintenance and/or lack of reinforcement, which is frequently detected not only in the expected case of informal urban settlements, but also in historical centres.

The high levels of vulnerability and occupancy of adobe constructions, together with the moderate-to-high seismic hazard of regions where they are located, motivate the huge economic and human losses recorded in past events. Emblematic disasters are those observed in recent earthquakes occurred in Northridge (California, 1994), El Salvador (2001), Peru (1966, 1970, 2001, 2007, 2018), Iran (2003), Afghanistan (2005), China (2008 and 2009), Chile (1985, 1995, 2010, 2015) and Pakistan (2015). Out of 1996 destructive earthquakes recorded since 1900–2011, 38% of total death toll was caused by the collapse of adobe and rubble masonry structures (Daniell and Vervaeck 2012). After the 1994 Northridge earthquake (Mw 6.7, 57 dead), Webster and Tolles (Webster and Tolles 2000) inspected 29 historic adobe buildings, concluding that out-of-plane failure modes of those constructions may initiate under peak ground accelerations between 0.1 and 0.2 g. A damage assessment after the 2001 El Salvador earthquake (Mw 7.7, 952 dead) highlighted heavy damage or collapse of more than 150,000 adobe houses (Dowling 2004). Almost 25,000 adobe dwellings collapsed and 36,000 adobe dwellings were severely damaged by the 2001 Peru earthquake (Mw 8.4, 81 dead) (Blondet et al. 2003). In 2003, the Bam earthquake (Mw 6.6, 26,000 + dead) caused the collapse of a huge amount of adobe houses, resulting in more than 60,000 homeless. Damage assessments after the 2007 Pisco earthquake in Peru (Mw 7.9, 519 dead) confirmed that outward overturning of façade walls is one of the most common failure modes suffered by adobe buildings (Blondet et al. 2011). Damage inspections after the 2015 Hindu Kush earthquake in Afghanistan (Mw 7.5, 399 dead) evidenced some form of damage to 70–80% of the adobe buildings located in two districts of Pakistan (Ismail and Khattak 2016). Damage consisted mainly in the out-of-plane collapse of adobe walls and failure of building corners, and to a lesser extent, in the flexural cracking of spandrels associated with in-plane loading of masonry walls with openings.

The seismic capacity of unreinforced earthen buildings is generally one order of magnitude less than those masonries with goods mortars that can withstand earthquakes. The advantage to having seismic reinforcement in earthen houses is that their displacements and ductility increases to the effect of having reinforcements (using materials compatible with the earth, that have capacities to resist strong tractions and adequate elasticity) (Blondet et al. 2016).

In the case of Peru, a high seismic activity country, from its prehistory there are pieces of evidence of internal reinforcement in earthen walls (e.g. Caral). Figure 7 shows an area in Caral placed on the top of a pyramid. The vertical timber reinforcement inside the adobe walls worked to make the walls more flexible during earthquakes and to withstand the vertical weight of the roof.

Caral's experience (trial-error) led its builders to use bags of straw rope meshes filled with stones at the bottom part of the nucleus or core of the stepped pyramids. This foundation system is named Shicras system isolation and served to dissipate seismic energy during earthquakes. They lasted 5000 years and left us clues to understand the seismic dissipation. Nowadays, the Shicras are the basis of ongoing research in Peru to create a natural seismic isolation systems for new adobe or earthen buildings. The seismic isolation proved consists of three layers (two stone layers and one wooden layer) inside of a rope bag. The configuration is as follows (Fig. 8):

(1) Lower layer of river stones with an average height from 0.10 to 0.15 m.
(2) Intermediate layer of ½" to ¾" river stones with an average height from 0.06 to 0.08 m. This layer controls the displacement of the lower layer and also dissipate seismic energy.
(3) Upper layer, a wooden collar beam that prevents differentials settlements.

In summary, in today's practice, to improve the seismic performance of adobe constructions, the adobe walls need to be a reinforcement, internal or externally. For example, walls are wrapped with synthetic meshes (geogrids) or knotted synthetic

Fig. 7 Original reinforcements of half-timbered, reed, and vegetable fibers in the earth walls of Caral built 5000 years ago, similar to the Quincha technique of the eighteenth century (Quincha means "wall" in the Quechua language)

Fig. 8 Stone seismic isolation for earthen houses

rope meshes (Blondet et al. 2011). These meshes give a controlled displacement and some ductility to the adobe masonry, and also controlling the widening of the cracks and allowing some energy dissipation during the earthquake, avoiding the partial or total collapse of the structure. These reinforcement systems are recommended in the Adobe Peruvian Standards (NTE E080 2017).

Recently in the highland city of Orduña (Puno, Peru), the PUPC's academy developed a three-year project to improve the seismic performance of the typical adobe houses and also to solve the thermal problems. Orduña is a place located close to the highest lake in the world: Lake Titicaca, around 5000 m over the sea level, and its population suffer annual frosts. The healthy and safe home program ended in mid-2016. Studies of how to solve thermal problems using natural materials (e.g. totora) were combined with seismic reinforcements based on synthetic halyard meshes accessible in Puno. On December 1, 2016, an Ml 6.3 occurred in Orduña. Unfortunately, many unreinforced adobe houses collapsed (Fig. 9) while the reinforced ones were able to withstand the movement showing very few and slight cracks (Fig. 10). Also, stone masonry houses reinforced within the pilot project had a good seismic behavior (Fig. 11).

Fig. 9 Effect of earthquakes collapsed unreinforced adobe buildings after the Orduña's earthquake

Fig. 10 Reinforcement adobe houses with synthetic halyards meshes and plastered with earth and straw

Fig. 11 Stonemasonry house with earthen mortar, reinforced with synthetic halyard meshes

4 Contribution of Institutions for the Conservation of Buildings with Cultural Value

The ICOMOS International Council of Monuments and Sites is the institution in charge of defining and transmitting the knowledge to maintain, restore, and reinforce the buildings that have a Cultural legacy. ICOMOS does it through Conservation Letters, Doctrinal Texts, and Declarations of Cultural Preservation.

From the twentieth century, ICOMOS documents establish some engineering aspects related to the conservation of historical buildings, as in the Victoria Falls Charter (2003) and the Lima Declaration (2010). It is important also to highlight the research efforts from other institutions as the Getty Conservation Institute (GCI) within its Getty Seismic Adobe Project (GSAP 1990–96). The GCI organized training courses at the postgraduate level, conducting also international events and dissemination on the topic of the structural analysis of historical constructions.

To address the issue of the conservation of cultural heritage, recently ICOMOS has been adapting and developing some documents, and at the same time, the academy has tackled new fields of research on topics such as material assessment, the study of the constructive forms, numerical modeling, structural assessments and diagnosis, seismic vulnerability, etc. Many of these developments are due to the new knowledge to protect and build better seismic-resistant buildings and maybe also used in the task of conserving the built heritage in areas periodically altered by other natural disasters. These last issues were not covered in the ICOMOS Conservation Letters that were born in the twentieth century. However, the World Heritage Center (WHC), a UNESCO office that safeguards the world heritage, recommends that the ICOMOS National Councils establish preservation solutions that respond to the ecology and reality of each country.

There are already national guidelines in many countries that frequently suffers from natural disasters, and it will undoubtedly be the new development in this century.

5 Conclusions

Adobe construction goes back around 9000 years. The first civilization to use moulded adobes was located in Mesopotamia, and the first one to use seismic improvements in their earthen constructions was placed in Caral (Peru) 5000 years ago. The earthen constructions do not have an excellent structural behaviour during earthquakes, they are massive and easily break under dynamic actions. Then, the need for seismic reinforcement is necessary. Nowadays some national construction standards specify methodologies for seismic strengthening of vernacular earthen dwellings.

The catastrophic consequences of natural events (particularly earthquakes and floods) in regions with huge amounts of adobe constructions are a tangible evidence of an urgent call for risk reduction programmes (at local and urban scale). The latter become a rather complex task for the following reasons: (i) many adobe constructions are located in less developed countries where structural strengthening is feasible only in case of low-cost and low-tech solutions; and (ii) strengthening systems for cultural heritage sites and ancient constructions should meet environmental sustainability requirements and restoration principles. The latter include the physical-mechanical compatibility between strengthening and existing materials, the architectural-historical compatibility with the existing constructions, and reversibility.

Therefore, there is the need to study the mechanical behaviour of the adobe material and to know the structural behaviour of adobe buildings: vernacular and historical ones. In this last, it is also important to understand the different ways for improving its performance under the effect of natural phenomena, as floods, earthquake, etc. In the following chapters, a better view of different researches carried out considering earthen structures around the World is showed.

References

Achenza M, Sanna U (eds) (2009) Il manuale tematico della terra cruda. DEI, Rome, Italy
Avrami E, Guillaud H, Hardy M (eds) (2008) Terra literature review—an overview of research in earthen architecture conservation. The Getty Conservation Institute, Los Angeles, California, USA
Baglioni E, Fratini F, Rovero L (2010) The material utilised in the earthen buildings sited in the Drâa Valley (Morocco): Mineral and mechanical characteristics. In: Proceedings of 6th seminar of earthen architecture in Portugal and 9th Ibero-American seminar on earthen construction and architecture (CD-ROM). Center for Archaeological Studies at the Universities of Coimbra and Porto, Coimbra, Portugal
Berge B (2009) The ecology of building materials, 2nd ed. Architectural Press, Elsevier Science
Blondet M, Villa-García G, Brzev S (2003) Earthquake-resistant construction of adobe buildings—a tutorial. EERI/IAEE World Housing Encyclopedia, Okland, California, USA
Blondet M, Vargas J, Tarque N, Iwaki C (2011) Construcción Sismorresistente en tierra: la gran experiencia contemporánea de la Pontificia Universidad Católica del Perú (in Spanish). J Informes de la Construcción 63(523):41–50
Blondet M, Vargas J, Tarque N, Sosa C, Sarmiento J (2016) Seismic protection of earthen vernacular and historical constructions. In: Proceedings of the 10th international conference on structural analysis of historical constructions, SAHC
Correia M et al (2011) Terra Europae—earthen architecture in the European union. Edizioni ETS, Pisa, Italy
Correia M (2016) Conservation in earthen heritage—assessment and significance of failure, criteria, conservation theory, and strategies. Cambridge Scholars Publishing, UK
Daniell JE, Vervaeck A (2012) CEDIM earthquake loss estimation. Series research report 2012-01, CEDIM, Karlsruhe, Germany
Delgado MCJ, Guerrero IC (2006) Earth building in Spain. Constr Build Mater 120:679–690
Dell'Acqua AC, Franzoni E, Sandrolini F, Varum H (2009) Materials and techniques of art Nouveau architecture in Italy and Portugal: a first insight for an European route to consistent restoration. Int J Restor Build Monuments 15(2):129–143
Dowling D (2004) Adobe housing in El Salvador: earthquake performance and seismic improvement. In: Rose I, Bommer JJ, López DL, Carr MJ, Major JJ (eds) Natural hazards in El Salvador. Geological Society of America Special Papers, pp 281–300
Duarte I, Pedro E, Varum H, Mirão J, Pinho A (2017) Soil mineralogical composition effects on the durability of adobe blocks from the Huambo region, Angola. Bull Eng Geol Environ 76:125–132
Fernandes M, Portugal MV (2006) Atlântico versus Portugal Mediterrâneo: Tipologias arquitectónicas em terra. In: International conference Terra Brasil 2006, Ouro Preto, Minas Gerais, Brazil
Fratini F, Pecchioni E, Rovero L, Tonietti U (2011) The earth in the architecture of the historical centre of Lamezia Terme (Italy): characterization for restoration. Constr Build Mater 53: 509–516
Houben H, Guillaud H (1994) Earth construction—a comprehensive guide. Intermediate Technology Publication, London, UK
Ismail N, Khattak N (2016) Building typologies prevalent in Northern Pakistan and their performance during the 2015 Hindu Kush earthquake. Earthq Spectra 32(4):2473–2493
Minke G (2006) Building with earth—design and technology of a sustainable architecture. Birkhäuser, Basel, Switzerland
Mousavi SE, Khosravifar A, Bakhshi A, Taheri A, Bozorgnia Y (2006) Structural typology of traditional houses in iran based on their seismic behaviour. In: Proceedings of 8th U.S. national conference on earthquake engineering, San Francisco, California, USA
NTE E080 (2017) Diseño y Construcción con Tierra Reforzada. Ministerio de Vivienda, Construcción y Saneamiento

Parisi F, Asprone D, Fenu L, Prota A (2015) Experimental characterization of Italian composite adobe bricks reinforced with straw fibers. Compos Struct 122:300–307

Sanna A, Atzeni C (2009) Architettura in terra cruda. DEI, Rome, Italy

Tanaçan L (2008) Adobe construction: a case study in Turkey. Architectural Sci Rev 51(4): 349–359

Tavares A, Costa A, Varum H (2012) Adobe and modernism in Ílhavo, Portugal. Int J Architect Heritage 6(5):525–541

Varum H, Figueiredo A, Silveira D, Martins T, Costa A (2011) Outputs from the research developed at the University of Aveiro regarding the mechanical characterization of existing adobe constructions in Portugal. Informes de la Construcción 63(523):127–142

Webster F, Tolles L (2000) Earthquake damage to historic and older adobe buildings during the 1994 Northridge, California Earthquake. In: Proceedings of 12th world conference on earthquake engineering, Auckland, New Zealand

World Heritage Earthen Architecture Programme (WHEAP) (2012) World heritage inventory of earthen architecture. CRATERRE-ENSAG

Wu F, Li G, Li H-N, Jia J-Q (2013) Strength and stress-strain characteristics of traditional adobe block and masonry. Mater Struct 46(9):1449–1457

Behaviour of Adobe Construction in Recent Earthquakes

Nicola Tarque, Erkut Sayın, Muhammad Masood Rafi, and E. Leroy Tolles

Abstract The adobe masonry is classified as a quasi-brittle material. This is because the material fails under very low tensile stresses. Then, under the action of any type of loading (especially dynamic one), adobe masonry behaviour rapidly changes into nonlinear behaviour. However, adobe masonry resists moderate compressional loading. During earthquake actions, the adobe material starts to fail at the zones of stress concentration, such as corners of openings. Also, vertical cracks at the intersection of two orthogonal walls may appear. This is due to the absence of confinement elements that could guarantee a box behaviour on each floor. If walls continue breaking, then the most probable failure is due to the overturning of walls and the roof collapse. In this chapter, the most common types of failure of adobe buildings are shown and discussed based on field surveys carried out after some earthquakes.

Keywords Adobe masonry · Earthen constructions · In-plane failure · Out-of-plane failure

N. Tarque (✉)
GERDIS Research Group, Civil Engineering Division, Department of Engineering, Pontificia Universidad Católica del Perú, Av. Universitaria 1801, San Miguel, Lima, Peru
e-mail: sntarque@pucp.edu.pe

E. Sayın
Civil Engineering Department, Faculty of Engineering, Firat University, 23119 Elazığ, Turkey
e-mail: erkutsayin@gmail.com

M. M. Rafi
Department of Earthquake Engineering, NED University of Engineering and Technology, Karachi 75250, Pakistan
e-mail: rafi-m@neduet.edu.pk

E. L. Tolles
ELT & Associates, Oakland, CA 94610, USA
e-mail: rtolles@44adobe.com

H. Varum et al. (eds.), *Structural Characterization and Seismic Retrofitting of Adobe Constructions*, Building Pathology and Rehabilitation 20,
https://doi.org/10.1007/978-3-030-74737-4_2

1 Introduction

In many developing countries soil is still a widely used construction material because it is readily available at little or no cost. Most people in these countries, therefore, have no alternative but to build with soil, because the cost of manufactured or industrial materials such as wood, fired clay bricks, cement, or reinforcing steel is completely beyond their economic means.

Because building with earth is relatively simple, it is usually performed by the residents themselves, without technical assistance or quality control, resulting in houses with high seismic vulnerability. As a consequence, every significant earthquake that has occurred in regions where earthen construction is common has produced tragic life losses and considerable material damage. The most common type of failure of these constructions is the overturning of the walls, basically due to the lack of confinement and a rigid diaphragm that guarantee a box behaviour.

According to a literature review (Blondet et al. 2001; Dowling 2004), the typical failure of adobe constructions due to earthquakes may be summarized as follows (Fig. 1):

- Vertical cracking at the wall's corners due to the low tensile strength of the adobe material. Vertical cracks are the most typical cracks in earthen buildings.
- Diagonal cracking in walls due to in-plane forces, which generates tensile stresses at around 45°. These cracks start from the opening corners and go to the top and bottom part of the wall.
- Disintegration and overturning of walls at upper regions due to bending actions. Also, overturning of wall panel due to vertical cracks at the wall intersections.
- Roof collapse due to the overturning of pier walls.

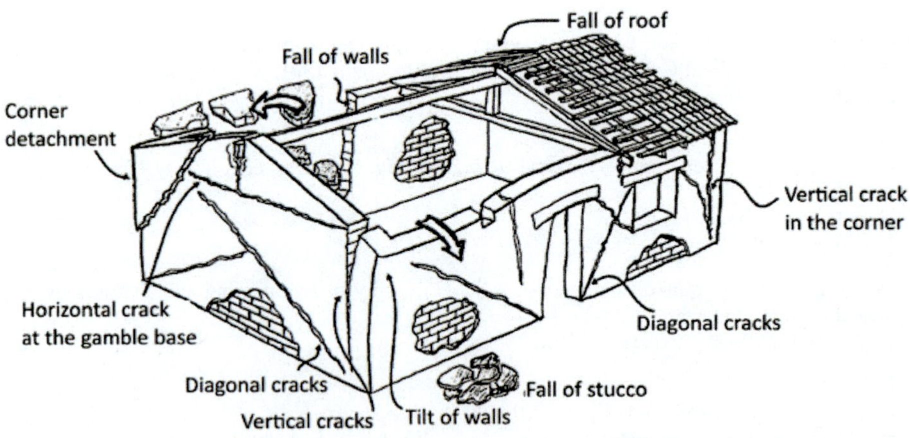

Fig. 1 Seismic deficiencies on adobe masonry (CENAPRED 2003)

2 Behaviour of Adobe Constructions During the Iran Earthquake, 2013

A strong earthquake (Mw 7.8) occurred 83 km East of Khash in Iran on 16 April 2013 (EERI: Earthquake Engineering Research Institute 2013) at a focal depth 82 km (USGS: United States Geological Survey 2013). This earthquake caused building damages in Iran, Pakistan and India. The town of Mashkel lied within 100 km radial distance from the epicentre of the earthquake in Balochistan, Pakistan. This was the nearest habituated town near the border of Iran. A large majority of housing units were made of adobe in this town and adjacent areas. The presence of adobe in this region is partly owing to its remoteness from any industrial city which makes the availability of modern construction materials difficult and costly. This was further complemented by the availability of suitable clay nearby.

A large number of adobe structures were severely damaged in Mashkel in the aftermath of the 2013 Iran earthquake. This includes an old adobe castle which (according to some reports) was built in 1889 by the then British Government. Its boundary wall and check posts collapsed at several locations. Extensive cracking was also observed in the walls. These damages were similar to those observed in other adobe structures in the region as discussed in the forthcoming sections.

2.1 *Out-of-Plane Wall Failure*

Different types of out-of-plane wall failures were observed in the damaged adobe buildings. Figure 2 illustrates an out-of-plane failure of a façade wall due to its overturning during the ground shaking. Figure 3 illustrates a failure mode after the crack at a wall junction was formed by a combination of shearing and tearing stresses. The crack extended up to the base of the out-of-plane wall in the vertical plane. The crack traversed in a horizontal direction after reaching the wall base. The out-of-plane movement of the wall caused its overturning. In other cases, wall cracking at the junction caused increased out-of-plane deformation of the wall. As a result, the walls pounded against each other at the junction and this impact caused the failure of the wall corner in the top region (Fig. 4). Figure 5 presents the wall failure by the formation of a free-standing slender column which was created at the wall junction. The column collapsed during the ground shaking.

Fig. 2 Failure of façade wall due to out-of-plane overturning

Fig. 3 Wall failure initiated by shearing and tearing stresses

Fig. 4 Corner failure due to wall punding

Fig. 5 Failure of a slender column at the wall corner

Fig. 6 In-plane wall damages

2.2 *In-Plane Wall Failure*

The observed in-plane wall damages in adobe structures comprised of rocking and diagonal wall cracking. Figure 6 shows a combination of diagonal cracking and wall rocking in a wall subjected to in-plane shear. Rocking represents rigid body rotation of the wall about its compression side along with a horizontal crack. Diagonal cracks develop in the region of high shear in the plane of the wall after the principal tensile stress exceeds the material tensile strength. Stepped cracks were also observed in the in-plane walls which formed due to weak mortar as compared to the adobe units (Fig. 7).

Fig. 7 Stepped cracks in in-plane wall

2.3 *Failure of Roof*

Thatched roofs were generally employed in adobe buildings in Mashkel. The roof was supported on a horizontal support system made of bamboos in the longer direction and date palm tree trunk (which is cut longitudinally in the centre) or bamboos in the transverse direction. The bamboos and tree trunk rested directly on the wall. Since the roof acted as flexible diaphragms, it was unable to provide an effective diaphragm action due to the absence of bond beams. The out-of-plane walls act as vertical cantilevers due to lack of connection with the roof. The observed roof failure mode indicated that the roof translated as a rigid body during ground shaking. An out-of-plane failure of the wall caused the roof to collapse as it was supported directly on the wall (Fig. 8). On the other hand, the failure of the roof was avoided in those buildings where the roof spanned the walls not collapsing out-of-plane (Fig. 9).

Fig. 8 Typical roof failure

Fig. 9 Roof supported in the shorter span

3 Behaviour of Adobe Constructions During the Maden-(Elazığ) Earthquake, 2011

Turkey is one of the earthquake-prone countries in the world because of its geographical location. The tectonic in and around Turkey depends on relative motions between the African, the Aegean, the Arabian, the Anatolian, the Black Sea and the Eurasian plates (Kasapoglu and Toksöz 1983). Three major structures form the neotectonics of Turkey; they are dextral North Anatolian Fault Zone (NAFZ), sinistral East Anatolian Fault Zone (EAFZ) and the Aegean–Cyprian Arc. Also, sinistral Dead Sea Fault Zone has an important role (Bozkurt 2001). Elazığ locates on the East Anatolian Fault (EAF), almost at the north-eastern end, where EAF meets North Anatolian Fault (NAF) at the Karlıova triple junction (Celep et al. 2011). An earthquake of Mw 6.1 struck Kovancılar county of Elazığ on March 8, 2010, at 2:32 GMT. Followed by the main event, an aftershock caused damages in Palu country of Elazığ at 07:47 GMT with Mw 5.5. The earthquake resulted in 42 deaths and 137 injured. In the earthquake, considerable structural damage and fatalities occurred in some villages of Kovancılar.

The building stock of the villages consists of masonry and adobe structures mainly. Adobe brick consists of clayey soil, water and straw. After these components are mixed and filled into moulds, dried under the sun. Nearly all the buildings of the region have only one or two stories. In the earthquake area, adobe buildings built by local people without any engineering knowledge is preferred due to the advantages such as thermal properties and local material availability. Main reasons for structural damages in the buildings during the earthquake are presented below.

Fig. 10 Out-of-plane mechanism

3.1 *Out-of-Plane Failure*

Out-of-plane collapse may occur from the combination of different deficiencies. The main causes of out-of-plane failure of walls and gable end walls are lack of bond beams, poor wall to floor connections, poor wall to wall connections and large unsupported wall lengths (Sayın et al. 2014). Due to the weak connections of the walls to the slab or the roof system, they behave as a cantilever because of supported only at the base. The whole or the significant parts of the wall fall during the earthquake. Additionally, wooden logs of the roof are aligned in one direction and the beams are supported by two walls opposite to each other. The weight of the heavy roof and corresponding horizontal earthquake loads are transferred to these walls (Celep et al. 2011). Therefore, the walls that are not supported by the wooden logs may easily overturn to out-of-plane direction during the earthquake (Fig. 10).

3.2 *In-Plane Failure*

When the seismic forces are acted parallel to the wall plane, they cause to generate shear forces. Because of earthquake loads, shear forces increase and can damage walls. These damages generally occur near openings because of the lack of bond beams that distribute the lateral forces uniformly. In-plane damage may be observed in three different forms. The damage arises from some factors such as axial compressive load level, wall aspect ratio, and quality of the mortar in components that comprises the wall. Depending on these factors, in-plane damage may be observed in the form of diagonal, stepped and horizontal cracks (Fig. 11). The diagonal crack becomes X-shaped crack when the earthquake is reversed.

Fig. 11 In-plane mechanism

3.3 *Failure of Roof*

Heavy earthen roofs are very prevalent in the region. The roofs are generally made from thick layers of dried clay or mud placed on wooden logs. These logs having weak support connections and small support lengths on the walls. Thick roofs are preferred because of heat and water insulation properties. These earthen roofs lose their impermeability because of weather conditions. Therefore, they are filled with fresh soil in springtime and they are compacted with the use of heavy stone cylinder (Korkmaz et al. 2010). This cylinder is called a log. Thus, the weight and thickness of the roof increase over time. The thickness of the earth generally varies 0.30–0.60 m which generates large inertial forces under earthquake actions. Figure 12 shows the damaged masonry buildings arising from the heavy earthen roof.

Fig. 12 Heavy earthen roofs

Fig. 13 Corner damages

3.4 Corner Damages

Corner damages are common in the adobe buildings. Damage frequently occurs at the corner of buildings due to increasing stress concentrations during an earthquake. Thus, diagonal and vertical cracks occur in the corner of the building. Vertical cracks develop during the interaction of perpendicular walls and are caused by flexure and tension due to out-of-plane movements. Also, in-plane shear forces cause diagonal cracks that start at the top of a wall and extend downward to the corner (Fig. 13). This type of crack results in a wall section that can move laterally and downward during the earthquakes (Adobe: Seismic Retrofitting Guidelines of Buildings in Nepal 2016).

4 Behaviour of Adobe Constructions During the Pisco Earthquake, 2007

The majority of seismic events that occur on the Peruvian coast are produced by the subduction process, where the Nazca plate moves under the South American plate at a rate of 80 mm per year. The earthquake that took place on August 15 (Mw 7.9) is related to previous seismic activity recorded in the area, such as the one that occurred on October 20, 2006. The 2006 earthquake had a Mw 6.4 magnitude and its hypocentre was located 90 km west of Pisco, at a depth of 43 km. The rupture area of both earthquakes took place on a seismic gap previously registered between the rupture areas of the 1974 Lima earthquake (Mw 7.6) and the 1996 Nazca earthquake (Mw 7.7).

As observed in all the affected cities around Pisco (a southern Peruvian city), the collapse in earthen constructions was triggered by the progressive formation of cracks in the walls. The most common types were vertical cracks at the wall's corners and the x-shaped cracks in the façade walls. The cracking of the walls triggered the collapse of the structure and the partial or complete overturning of the walls caused the roof's collapse. The effect that the earthquake had on adobe constructions was catastrophic. In the Pisco area, for example, nearly 80% of the adobe buildings were completely damaged and destroyed by the earthquake. The high collapse percentage on the adobe buildings can be attributed mainly to the lack of seismic reinforcement on the dwellings. Other factors that influenced the buildings' collapse and that were responsible for much of the structural damage were the soft soils, the low quality of the building materials and the labour, the thinness of the walls, the inadequate configuration and location of openings (doors and windows), and the weak bond in the intersection between the walls and the roof.

4.1 Out-of-Plane Failure

The most common failure observed in earthen buildings, especially in those with thin walls, was the overturning of the façade walls and their collapse onto the street. This was caused because the wall strength in the intersection between the façade wall and the other house walls was too low to withstand the earthquake's movement. The walls usually collapsed as follows: first vertical cracks appeared on the wall's corners causing the adobe blocks in that area start to break and fall (Fig. 14a). This triggered the walls to disconnect until finally, the façade wall turned over (Fig. 14b). The situation was worsened by the fact that separation joints between the buildings seem to be an uncommon practice in the area. Consequently, the direct contact between the walls during seismic activity leads to their collapse. The observations made after the earthquake have shown that the magnitude of the damage that the buildings suffered when their walls collapsed, was directly related to whether the roof's wooden joists where supported on the façade wall or not. If they were supported by the façade, the wall's collapse caused them to come off

(a) Vertical cracks in the wall's corner (b) Collapse of the façade walls

Fig. 14 Damage on adobe buildings due to the Pisco earthquake

Fig. 15 Diagonal and vertical cracks shown on corner buildings

balance, causing the roof to collapse as well. If, on the other hand, the joists were supported by the walls that were perpendicular to the façade wall, the roof didn't fall apart (Fig. 14b).

Many adobe buildings located on street corners suffered from heavy structural damage due to the collapse of their two façade walls and roof. The cracks produced on the corner between these two walls are both vertical and diagonal. The diagonal cracks extend from the highest part of each wall's corner down to the house base, forming a 'v' like crack pattern (Fig. 15).

4.2 *In-Plane Failure*

Lateral seismic forces acting within the plane of the walls generate shear forces that produce diagonal cracks, which usually follow stepped patterns along the mortar joints. The diagonal cracks often start at the corners of opening doors and windows, due to the stress concentration at these locations (Fig. 16). If the seismic movement

Fig. 16 Typical X-shape cracks on adobe walls due to in-plane actions

continues after the adobe walls have cracked, the wall breaks in separate pieces, which may collapse independently in an out-of-plane fashion.

4.3 *Quality of the Constructions Materials*

The adhesion between the adobes and mortar in most collapsed or heavily damaged walls (especially in the ones belonging to the more recent buildings in Pisco) was quite poor and weak. When the adobe walls collapsed, most of the adobe blocks that were used for these constructions didn't hold together with the mortar and crumbled (Fig. 17). It is presumed that, as it is obtained from a coastal area, the soil used for the making of the bricks and mortar contains too much sand and too little clay, producing, as a result, a weak bond between the mortar and the adobe blocks. Unfortunately, clay is the only soil component that, as a reaction to water contact, causes the soil particles to chemically bond. In some of the houses in the area, it was observed that the adobe used for the constructions would easily crumble when scratched with the nail.

4.4 *Construction on Developing Areas*

Some of the more recent buildings located on the developing areas around Pisco have been constructed without considering seismic security guidelines and are, therefore, more susceptible to suffer damage due to seismic activity than the older constructions in the area. The walls on the modern buildings are more susceptible to collapse due to out of plane loads as they are narrower (with a 0.25 m thickness) and slenderer than the older walls. Many of the houses in the area only have 1 or 2 enclosed rooms, and many of them have enclosure walls constructed using adobe. The roofs were built using wood that was covered with a crushed cane, straw mats,

Fig. 17 These images show the weak bond between the mortar and the adobe blocks

Fig. 18 Collapse of buildings located on developing urban areas

cardboards or sandbags, used as insulation to keep the house inhabitants warm. The floor of the houses is usually moistened soil that has been compacted. Most of the dwellings constructed following this model collapsed during the earthquake (Fig. 18) basically by overturning of the walls.

4.5 Damage in Earthen Historical Landmarks

Most of the earthen churches in the Peruvian coastal area have one or two towers that are about 3–4 story high as part of their façade, lateral adobe walls (1.20 m thick and 5–6 m high approximately), and a roof consisting of a cylindrical vault constructed using wooden arches, crushed cane, straw mats and a mud coat on the outside and stucco on the inside (Fig. 19). The plan view of a typical church is cross-shaped. The choir area (which is 2-story high) is located right behind the church's main door and the altar is on the back of the church. The church's lateral

(a) Front view (b) Internal view

Fig. 19 Typical earthen churches at the Peruvian coastal area walls have pilasters built approximately every 4 m that simulate columns (Fig. 19b).

Almost all of the churches and historical landmarks that had been built with soil and bricks suffered heavy damage or even collapsed as a consequence of the earthquake. According to the National Institute of Culture (INC), 32% of the historical and cultural monuments around Pisco city have completely collapsed, 23% were under strong risk of collapsing, 26% were under moderate risk and 19% showed minor damage to their structure (Libre 2003). The most common failures observed in the local churches were:

- Horizontal cracks on the lateral walls at about 1/3 of their total height (Fig. 20a). These cracks can even break through the earthen pilasters (Fig. 20b), causing the walls to collapse.
- Diagonal cracks on some of the lateral walls.
- A detachment of the choir (Fig. 20c) and the altar's wall (parallel to the façade) from the church's lateral walls and cylindrical vault ceiling (Fig. 20a).
- A detachment of the towers from the rest of the church (Fig. 20d).
- The appearance of vertical cracks and fissures on the church towers (Fig. 20d).
- Humidity related damage.

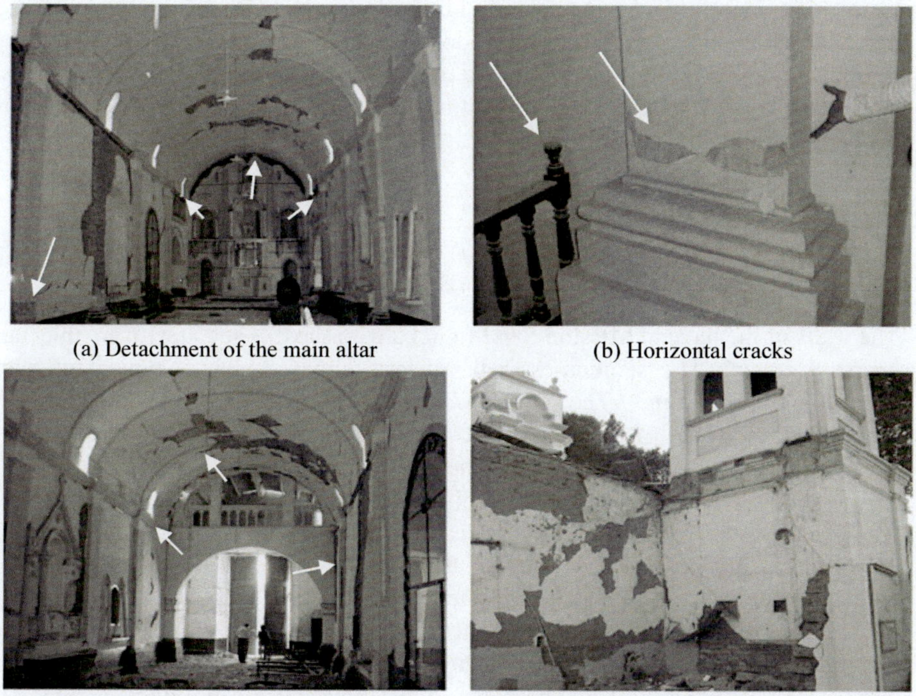

(a) Detachment of the main altar

(b) Horizontal cracks

(c) Detachment of the choir from the vault

(d) Vertical cracks on theintersection between the tower and walls

Fig. 20 Most common earthquake damages on earthen churches

5 Behaviour of a Historic Adobe Church During the San Simeon Earthquake in California, 2003

The Mission at San Miguel suffered earthquake damage during the San Simeon earthquake of 2003. A strong earthquake of magnitude 6.5 (Mw) struck the Central Coast of California on Monday, December 22, 2003. The epicenter was 11 km (7 miles) NE of San Simeon, at a depth of about 8 km (5 miles). The area where the quake struck displays complex faulting, between the Oceanic Fault and Nacimiento Fault zones, along with possible interaction from the Hosgri fault and San Simeon fault zones. The earthquake was probably caused by thrust faulting and the rupture propagated southeast from the hypocenter for 12 miles (19 km) (McLaren et al. 2003).

The area around the epicentre being sparsely populated, the most severe damage was in Paso Robles, 24 miles (39 km) east-southeast. Two women were killed when the Acorn Building, an unreinforced masonry structure built in 1892, collapsed. Other unreinforced masonry buildings, some more than a century old, were extensively damaged. However, none of the collapsed building had even partial retrofitting. At least 40 people were reported to be injured in the Paso Robles-Templeton area.

The observed damaged to Mission San Miguel is not a typical type of damage to many adobe buildings. Most adobe buildings do not have such massive long walls that are buttressed by smaller, thinner adobe walls. Much of the type of damage has been documented in this paper. Separately, typical damage was also documented in a report by the Getty Conservation Institute of earthquake damage to mostly historic adobe buildings after the Northridge Earthquake in 1994 (Tolles et al. 1996).

5.1 *Out-of-Plane Failure*

The walls of the nave at Mission San Miguel are massive. They are 1.6 m. thick and nearly 9 m. tall. The slenderness ratio is less than 6 which is typical of many historic adobe buildings. The church was constructed in the early 1800s. On the other hand, the walls of the sacristy are only 0.6 m thick which makes them quite slender and not nearly as massive as the walls of the nave.

These massive walls rock considerably during a larger earthquake and generate a significant amount of force. In this case, the long walls of the nave rocked out-of-plane and pushed the walls of the sacristy causing considerable in-plane damage. The crack at the interface of these two walls is 2 inches (5.08 cm) wide at the top and narrow to nothing at the bottom of the wall. The sacristy wall has significant in-plane crack damage exhibited by the cracks emanating from the corners of the window. Two diagrams of the building are provided in Figs. 21. Figure 21a is an overview of the entire building and Fig. 21b is a section through the building with arrows indicating the direction of rocking of the massive walls of the nave and the crack damage to the sacristy wall.

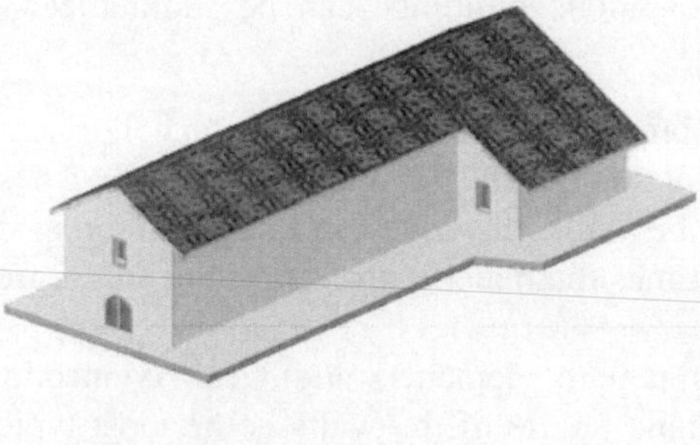

(a) Drawing of the building

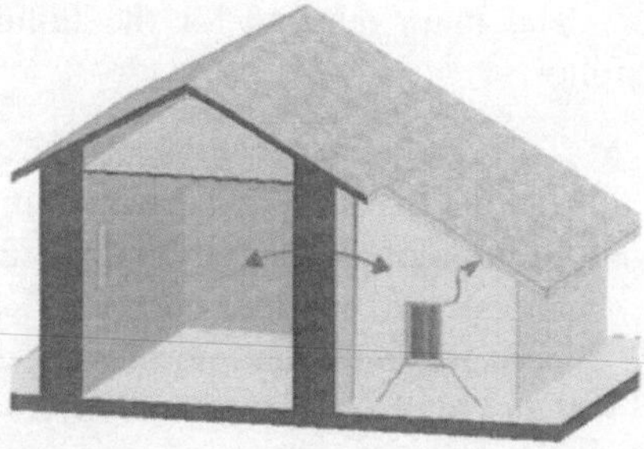

(b) Out-of-plane rocking of Nave walls causes in-plane damage to the Sacristy wall

Fig. 21 Mission San Miguel in California

The lesson to be learned here is that out-of-plane motions of massive walls can exert very large forces on walls that buttress them. The forces generated by the massive walls of the nave can overpower the strength of the much thinner and less massive walls of the sacristy.

Construction joints that provided gaps that allow the massive walls to rock out-of-plane can be helpful in reducing this type of damage. If the crack at the sacristy wall is filled every time rocking creates this type of opening, eventually, the rocking of the massive walls eventually will cause the collapse of the buttressing walls of the sacristy.

5.2 *In-Plane Failure*

The large forces created by the out-of-plane walls resulted in shear forces that produce diagonal cracks that extended outwards from 3 of the 4 corners on the opening in the wall of the Sacristy. These cracks are typical of in-plane failure in adobe walls. The exception is that no diagonal crack occurred at the upper left corner of the window because the forces were primarily delivered at the top left of the cracked, Sacristy wall from the rocking action of the wall of the Nave.

6 Concluding Remark

In this chapter, failure of adobe constructions (basically dwellings) in different parts of the world are investigated and commented based on field survey carried out after some important earthquakes. Although there are different typology constructions, it seems that the building failure is common in all of them basically due to the low tensional strength of the adobe masonry.

The main reasons for the failures of adobe buildings can be summarized as follows:

- The adobe material is brittle and breaks immediately after some horizontal actions that may exceed the tensional strength of the material.
- In most of the heavily damaged adobe walls, the adhesion between mortar and adobe units is quite poor, and sometimes the material used for units has a great percentage of sand and less of clay.
- Due to the lack of integrity (confinement elements) and weak connections among walls and roofs, an out-of-plane failure of the walls is the most typical failure during earthquakes. Vertical cracks suddenly appear at the wall corners and disconnect them. Then, after the walls collapse out-of-plane, the roof may also collapse.
- Out-of-plane failure may occur from the combination of different deficiencies in the damaged adobe buildings. Lack of bond beams, weak wall to floor connections, weak wall to wall connections and long unsupported wall length are some of these deficiencies.
- In-plane wall failure can be observed as the form of horizontal cracks indicating rocking and diagonal cracks indicating high shear where the principal stress exceeds the material tensile strength. Also, stepped cracks can be observed due to weak mortar as compared to the adobe units.
- In some cases, heavy earthen roofs are one of the reasons for the damage. These roofs generally place on wooden logs with weak support and small support lengths on the walls. During an earthquake, the walls cannot adequately support the heavy mass and the heavy roof partially or completely collapsed.
- In historic adobe buildings, the collapse of the gable wall is common and disconnection of different building parts appear, like a disconnection between towers and central walls parts in case of churches.
- Poor quality of the structural materials, poor workmanship during the construction and climatic conditions are another reason for adobe building damage.

Authors believe that an adequate retrofitting system that guarantees a box behaviour of the building (roof and walls) may reduce the probability of building collapse during earthquakes.

References

Adobe: Seismic Retrofitting Guidelines of Buildings in Nepal (2016) Government of Nepal Ministry of Urban Development, Kathmandu

Blondet M, Vargas J, Tarque N, Iwaki C (2001) Seismic resistant earthen construction: the contemporary experience at the Pontificia Universidad Católica del Perú. Informes de la Construcción 63(523):41–50

Bozkurt E (2001) Neotectonics of Turkey—a synthesis. Geodin Acta 14:3–30

Celep Z, Erken A, Taskin B, Ilki A (2011) Failures of masonry and concrete buildings during the March 8, 2010 Kovancılar and Palu (Elazığ) Earthquakes in Turkey. Eng Fail Anal 18(3):868–889

CENAPRED (2003) Reinforcement methods for self-construction of rural housing. Mexico City, Mexico (in Spanish)

Dowling D (2004) Adobe housing in El Salvador: earthquake performance and seismic improvement. In: Rose I, Bommer JJ, López DL, Carr MJ, Major JJ (eds) Geological society of America special papers, pp 281–300

EERI: Earthquake Engineering Research Institute (2013) M7.8 Earthquake strikes Iran. https://www.eeri.org/2013/04/m7-8-earthquake-strikes-iran/ Accessed on 25 Apr 2013

Kasapoglu EK, Toksöz MN (1983) Tectonic consequences of the collision of the Arabian and Eurasian plates: finite element models. Tectonophysics 100:71–95

Korkmaz HH, Korkmaz SH, Donduren MS (2010) Earthquake hazard and damage on traditional rural structures in Turkey. Nat Hazards Earth Syst Sci 10:605–622

Libre H (2003) El 32% de los monumentos en Ica se han perdido, afirma el INC, 2003. http://www.24horaslibre.com/nacionales/1188483202.php (in Spanish)

McLaren MK, Hardebeck JL, van der Elst N, Unruh JR, Bawden GW, Blair JL (2008) Complex faulting associated with the 22 December 2003 Mw 6.5 San Simeon, California, Earthquake, Aftershocks, and Postseismic Surface Deformation. Bull Seismol Soc America Seismol Soc America 98(4):1659–1680

Sayın E, Yön B, Calayır Y, Gör M (2014) Construction failures of masonry and adobe buildings during the 2011 Van earthquakes in Turkey. Struct Eng Mech 51(3):503–518

Tolles EL, Webster FA, Crosby A, Kimbro E (1996) Survey of damage to historic adobe buildings after the 1994 Northridge earthquake. GCI Scientific Program Report, Getty Conservation Institute

USGS: United States Geological Survey (2013) Significant earthquake and news headlines archive: Magnitude 7.8 - Iran-Pakistan Border Region. http://earthquake.usgs.gov/earthquakes/eqinthenews/2013/usb000g7x7/ Accessed on 25 Apr 2013

Mechanical Characterization of Adobe Bricks

Dora Silveira, Cristina Oliveira, Humberto Varum, Ioannis Ioannou, Lorenzo Miccoli, Nicola Tarque, Fulvio Parisi, Luigi Fenu, Mario Solís, and José D. Rodríguez-Mariscal

Abstract The mechanical characterization of adobe bricks is an important first step in the study of the behaviour of adobe masonry. For this reason, in the last decades, different authors have conducted research on the mechanical behaviour of adobes from various regions of the world. Despite the importance of mechanical characterization, there are still only a few standards and normative documents with clear indications for the mechanical testing of earthen materials and, in general, these indications are not thorough and vary among different countries. Consequently, authors tend to adopt different types of test specimens and procedures in their experimental work, thus obtaining results that are not directly comparable. The fact that the materials and procedures traditionally used are also not standardized, varying greatly from region to region, also contributes to the difficulty of comparing results from different studies. This chapter presents a review of the indications

D. Silveira (✉)
ADAI-LAETA, Itecons—Instituto de Investigação e Desenvolvimento Tecnológico para a Construção, Energia Ambiente e Sustentabilidade, 3030-289 Coimbra, Portugal
e-mail: dora.silveira@itecons.uc.pt

C. Oliveira
CONSTRUCT-LESE, University of Porto, Technology School of Barreiro, Polytechnic Institute of Setubal, 2839-001 Lavradio, Portugal
e-mail: cristina.oliveira@estbarreiro.pt

H. Varum
CONSTRUCT-LESE, Civil Engineering Department, Faculty of Engineering, University of Porto, 4200-465 Porto, Portugal
e-mail: hvarum@fe.up.pt

I. Ioannou
Ledra and Building Materials Laboratories, Department of Civil and Environmental Engineering, School of Engineering of the University of Cyprus, 1678 Nicosia, Cyprus
e-mail: ioannis@ucy.ac.cy

L. Miccoli
Division Building Materials, Bundesanstalt für Materialforschung und—prüfung (BAM), Unter den Eichen 87, 12205 Berlin, Germany
e-mail: lorenzo.miccoli@bam.de

H. Varum et al. (eds.), *Structural Characterization and Seismic Retrofitting of Adobe Constructions*, Building Pathology and Rehabilitation 20,
https://doi.org/10.1007/978-3-030-74737-4_3

provided by codes, standards and other technical recommendations for the mechanical testing of adobe bricks, as well as a detailed review of procedures adopted, and results obtained by different authors regarding the mechanical characterization of traditional adobe bricks. This chapter focuses, in particular, on the behaviour of adobe bricks when subjected to simple compression. It provides an overview of the existing knowledge and identifies needs for future research and development.

Keywords Adobe · Adobe masonry · Mechanical behaviour · Test procedures · Standards

1 Introduction

Adobe bricks have been used in traditional construction since ancient times, as the material used (earth) is widely available and no special technology is required to produce them. The manufacture and construction methods and procedures used are based on empirical knowledge that, with the world development and industrialization, tend to disappear. Nowadays, this material is mostly used in developing countries. In developed countries, it is mainly used for restoration purposes. The mechanical characterization of adobe bricks is a subject that has been having some attention due to the lack of information on the subject and the need for it in the rehabilitation of heritage buildings.

N. Tarque
GERDIS Research Group, Civil Engineering Division, Department of Engineering, Pontificia Universidad Católica del Perú, Av. Universitaria 1801, San Miguel, Lima, Peru
e-mail: sntarque@pucp.edu.pe

F. Parisi
Department of Structures for Engineering and Architecture, University of Naples Federico II, Via Claudio 21, 80125 Naples, Italy
e-mail: fulvio.parisi@unina.it

L. Fenu
Dipartimento di Ingegneria Civile, Ambientale e Architettura, Università degli Studi di Cagliari, Via Santa Croce 59, 09123 Cagliari, Italy
e-mail: lfenu@unica.it

M. Solís · J. D. Rodríguez-Mariscal
Department of Continuous Media Mechanics and Theory of Structures, University of Seville, San Fernando, 4, 41004 Seville, Spain
e-mail: msolis@us.es

J. D. Rodríguez-Mariscal
e-mail: josedrm@us.es

Standardized testing methods for evaluating the mechanical behaviour of adobe bricks and formal technical guidance for the development and results analysis of those tests are missing. This chapter compiles the few existent and incomplete codes, standards and technical recommendations. Additionally, it presents the results obtained from different authors in relation to the analysis of distinct parameters, as compressive strength, stress-strain relationships, Young's modulus and Poisson's ratio.

2 Codes, Standards and Technical Recommendations

National and international normative documents and standards for the mechanical testing of adobe specimens exist in some countries, as presented in Table 1, for the compression test of adobe blocks. These usually describe the recommended testing procedures and prescribe permissible strength values. Nevertheless, there seems to be little consensus on the various tests procedures described in the aforementioned standards, and therefore any results obtained are not directly comparable. Furthermore, questions arise regarding, for example, the effect of specimen size and type, treatment of specimens prior to testing, platen restraint, application of correction factors, etc.

In fact, specimens can be specified to be cubes, rectangular prisms or cylinders. The test rate may differ either in strength applied along time or velocity of the test machine. In addition, the strength limit is also distinct in the different standards. It is clear that there is a lack of a uniform formal technical guidance regarding the testing of adobe specimens.

In general, in most standards, specific types of earthen materials are not properly addressed. Methods for evaluating the unconfined compressive strength of cohesive soils use a geomechanic analysis rather than a common buildings applications point of view (Illampas et al. 2014). In addition, unfired earth is generally not mentioned in the standards. When mentioned, the information is incomplete, lacking guidance in specimen size, mixture proportions, and curing procedures (Wosick et al. 2014).

Furthermore, the heterogeneity of earthen materials, traditions and techniques used across the world, lead to strong variations in almost every aspect. Nevertheless, it would be helpful to have general guidelines applicable to all types of soil and earthen materials used worldwide, also including the use of additives as fibres, minerals or synthetic ones. A consistent correction factor that could be applied to effectively compare the results obtained by different authors would be extremely convenient and important.

Table 1 Codes, standards and technical recommendations for the mechanical testing of adobe specimens

	Document	Specimens	Test rate	Other indications	Strength limit
Compression test	14.7.4 NMAC (2015)	• Adobe blocks • Length $\geq$ twice the width	3.45 MPa/min	• Specimens tested in the flat position	• Mean strength $\geq$ 2.07 MPa • No individual unit may have a strength of less than 1.72 MPa
	Norma E.080 (2017)	• 10 mm cube specimens	…	…	• Mean strength of 4 best specimens (out of 6) $\geq$ 1 MPa
	NZS 4298[a] (1998)	• Aspect ratio between 0.4 and 5.0 • 200 mm cube specimens recommended	…	• Specimens loaded in the same direction as in the wall	• Least of the individual results in the set >1.30 MPa (for samples with aspect ratio of 1; for other ratios the required result shall be $1.30 \times 0.7/k_a$[b])
	Pima County Standard (2012)	• Adobe blocks • Dried to constant weight (at 20 ±9 °C, relative humidity $\leq$ 50%)	$\leq$ 1.3 mm/min	• Specimens loaded in the same direction as in the wall	• Mean strength $\geq$ 2.07 MPa • No individual unit may have a strength of less than 1.72 MPa
	Australian Handbook (Walker 2002)	• Adobe blocks[c] or cylindrical specimens • Oven-dry or saturated surface dry condition or at other measured moisture content	1–5 mm/min or 9–42 MPa/min	• Unconfined strength obtained by applying an aspect ratio correction factor to the measured values	…

[a]Indications for 'standard grade earth construction'
[b]k_a is the ratio factor
[c]A whole specimen should be used; in case where specimen size cannot be accommodated, representative part-units cut from the whole may be used

3 Research on the Mechanical Behaviour of Adobe Bricks Under Simple Compression

3.1 Adobe Bricks Main Characteristics

In this chapter, 16 studies on adobe bricks mechanical characteristics were analysed and compared, listed in Table 2. Some studies used adobe blocks collected from existing traditional constructions, while, in others, new adobe bricks were produced specifically for the study conducted. The latter were made with traditional materials and production methods, either by local producers, in different regions of the country, or in the laboratory, while trying to reproduce the long-established traditional processes.

Table 2 presents, for each study, the location of the research conducted, the composition of the adobe bricks, their dimensions, condition (new or collected from an existing construction) and age. Old adobes (up to 4000–6000 years old) and new adobes were analysed and tested, with different mixtures or proportions of sand, clay, silt and gravel, and, in one, case, with air lime binder. Also, some researchers analysed the influence of natural fibers in the behaviour of the bricks.

As the mechanical behaviour of adobe bricks is affected by the size, shape, composition and manufacturing process of the specimens, different values are obtained by different authors. Table 3 summarizes the procedures and main results of simple compression tests on adobe specimens performed by the different authors. It states the type and aspect ratio of the specimens analysed, as well the type of deformation measurement used. The values obtained (range or mean value) for the compressive strength, strain at peak strength, Young's modulus, and Poisson's ratio are presented. In addition, it is also included information on the calculation method used for the modulus of elasticity.

3.2 Specimens and Testing Procedures

Not all authors provide detailed information on the specimens and testing procedures used to assess the behaviour of adobe under compression, even though more recent works tend to provide more complete information. A review of the specimens and testing procedures used by different authors, based on the analysis of the works presented in Table 3, is presented in the following subsections.

3.2.1 Specimens Geometry

In the 16 studies analysed (Table 3), the following specimens are used: cubes (50% of studies), square prisms (31% of studies), whole bricks (25% of studies), half or a quarter of bricks (13% of studies), cylinders (19% of studies), and reduced-scale

Table 2 Main characteristics of adobe bricks studied by different authors

References	Location	Adobe bricks			
		Composition	Dimensions	Condition	Age
Blondet and Vargas (1978)	Peru	Clayey soil and straw	L = 40 mm W = 20 mm H = 8 mm and L = 60 mm W = 30 mm H = 8 mm	New (produced for the study)	…
Rivera and Muñoz (2005)	Colombia	Inorganic clay of low plasticity	Not indicated	Collected from existing construction	≈ 300 years
Liberatore et al. (2006)	Italy	Silty sand	L = 380 mm W = 155 mm H = 165 mm	Collected from existing constructions	Not indicated
Quagliarini and Lenci (2010)	Italy	Mixture of soil (lean clay with sand), sand (well graded sand with gravel) and straw, in variable proportions	L = 230 mm W = 150 mm H = 130 mm	New (produced for the study)	…
Fratini et al. (2011)	Italy	Gravel clay	Not indicated	Collected from existing constructions	>70 years (produced until the 1st half of the twentieth century)
Vega et al. (2011)	Spain	Clay soil and straw (25% or 33%, by volume)	L = 250 mm W = 120 mm H = 100 mm	New (produced for the study)	…
Adorni et al. (2013)	Turkmenistan	Silty soil and straw (and possibly organic additives like animal dung)	L = 400 mm W = 400 mm H = 120 mm	Collected from existing constructions	≈ 2000 years
Silveira et al. (2013)	Portugal	Sandy soil and air lime binder	L = 440 mm W = 280 mm H = 120 mm	Collected from existing constructions	>70 years (produced until the 1st half of the twentieth century)

(continued)

Table 2 (continued)

References	Location	Adobe bricks			
		Composition	Dimensions	Condition	Age
Wu et al. (2013)	China	Soil with 40–55% clay-silt and 45–60% sand, and straw (0.5%) (by weight)	L = 200 mm W = 90 mm H = 50 mm	New (produced for the study)	…
Illampas (2013)	Cyprus	Lean clay Lean clay with sand	L = 450 mm W = 300 mm H = 50 mm	New (produced in different regions of the country)	…
Illampas et al. (2014)	Cyprus	Soil with 60–90% clay-silt, 8–25% sand and 1–15% gravel, and straw (15–40%) (by volume)	L = 450 mm W = 300 mm H = 50 mm	New (produced in different regions of the country)	…
Wosick (2014)	USA	Silty clayey sand and straw	L = 356 mm W = 254 mm H = 102 mm	New (produced for the study)	…
Parisi (2015)	Italy	Clayey/silty sand and straw (0.64%, by weight)	L = 400 mm W = 200 mm H = 100 mm	New (produced for the study)	…
Liberotti et al. (2016)	Turkey	"Mud" and natural fibers	L ≈ 120–970 mm W ≈ 200–300 mm H ≈ 60–80 mm	Collected from existing constructions	4000–6000 years
Aguilar et al. (2017)	Peru	Sandy loam soil	L = 320 mm W = 220 mm H = 120 mm	Collected from existing constructions	≈ 1100 years
Illampas (2017)	Cyprus	Lean clay soil and calcarenite sand (1:1), and straw (0–5%) (by weight)	L = 450 mm W = 300 mm H = 50 mm	New (produced for the study)	…
Rodríguez-Mariscal et al. (2018)	Spain	Not indicated (soils from riverbank of Guadalquivir)	L = 320 mm W = 160/80 mm H = 80/160 mm	New (produced for the study)	…

L length, *W* width, *H* height

Table 3 Procedures and main results of simple compression tests on adobe specimens performed by different authors

Reference	Specimens		Deformation measurement	f_c^a (MPa)	ε_{peak} (%)	E (MPa)	E calculation method	ν	σ versus ε curves?
	Type	Aspect ratio							
Blondet and Vargas (1978)	Cubes	1	…	1.27–1.62	…	…	…	…	No
Rivera and Muñoz (2005)	Bricks	Not indicated	…	3.04	…	…	…	…	No
Liberatore et al. (2006)	Bricks and half bricks	0.8–1.3	…	0.29–1.56	…	…	…	…	No
Quagliarini and Lenci (2010)	Blocks (bricks cut into 4 parts)	0.9	Relative displacement of testing platens	2.14–2.88	2.9–11.5	98–211	Not indicated	…	Yes
Fratini et al. (2011)	Cubes	1	Relative displacement of testing platens	0.20–0.78	≈ 0.65–5.6[b]	15–87	Secant between start and end of linear segment	…	Yes
Vega et al. (2011)	Bricks	0.8[c]	…	3.69/3.99[d]	…	…	…	…	No
		1.2[c]		2.19/1.13[d]					
Adorni et al. (2013)	Prisms	1.5–2.1	Displacement of the load cell[c]	0.24–1.33	0.72–1.07	55–289	Tangent at 60% of peak stress	…	Yes
Silveira et al. (2013)	Cylinders	≈ 1.8[f]	Directly on test specimens	0.23–1.02	0.024–0.127	7609–25000	Secant between start and end of linear segment	…	Yes
	Cubes	1		0.28–1.21	…	…	…	0.10	No

(continued)

Table 3 (continued)

Reference	Specimens		Deformation measurement	f_c^a (MPa)	ε_{peak} (%)	E (MPa)	E calculation method	ν	σ versus ε curves?
	Type	Aspect ratio							
Wu et al. (2013)	Bricks	4[h]	Relative displacement of testing platens	1.39–1.70	…	41–42	Initial tangent	…	Yes
Illampas et al. (2014)	Cylinders	1	Relative displacement of testing platens	0.76–1.41[i]	>10	11–92	Secant between 5 and 50% of peak stress	…	Yes
	Cubes	1		0.60–1.75[i]		…			Yes
	Prisms	0.5		1.45–3.31[i]					Yes
Wosick et al. (2014)	Half and quarter scale bricks	0.4	…	2.03–8.89	…	…	…	…	Yes
Parisi et al. (2015)	Cubes	1	Relative displacement of testing platens	1.08	1.35	145	Secant at 1/3 of peak stress[g,i]		Yes
Liberotti et al. (2016)	Cubes	1	Relative displacement of testing platens	0.18–5.0	…	17–667	Secant between start and end of linear segment		Yes
Aguilar et al. (2017)	Prisms	2.7	Displacement of the load cell	1.16	1.24	147	Secant between 30% and 60% of peak stress	…	Yes
	Prisms	≈ 2	Directly on test specimens[j]		0.32	801	Secant at 1/3 of peak stress[g]		Yes
Illampas et al. (2017)	Cubes	1	Relative displacement of testing platens	0.93–4.50	0.5–1	271–1239	Regression analysis between 5% and 30% of peak stress	…	Yes
	Cubes	1							Yes
				2.05–4.53		…	…		

(continued)

Table 3 (continued)

Reference	Specimens		Deformation measurement	f_c^a (MPa)	ε_{peak} (%)	E (MPa)	E calculation method	v	σ versus ε curves?
	Type	Aspect ratio							
Rodríguez-Mariscal et al. (2018)	Cubes	1	Displacement of the actuator	1.13	6	33	Secant between 1/3 and 2/3 of peak stress	…	Yes
	Prisms	2	Displacement of the actuator and directly on test specimens	1.06	1.3/0.6 [k]	132/557 [k]		…	Yes
	Cyinders	2		1.33	1.6/0.4 [k]	195/802 [k]		…	Yes

f_c compressive strength, ε_{peak} strain at peak stress, E modulus of elasticity, v Poisson's ratio, σ stress, ε strain
[a]Single values correspond to the mean strength
[b]Estimated from the stress-strain curves presented by Fratini et al. (2011)
[c]Some bricks were loaded on the upper face (aspect ratio = 0.8) and others on the lateral face (aspect ratio = 1.2)
[d]25% of straw/33% of straw
[e]The measured displacements were corrected by subtracting the deformation of the press
[f]Specimens were cut with an aspect ratio of approximately 2, whenever possible, and never less than 1. For specimens with an aspect ratio ≤ 1.75, correction factors were applied in the calculation of the compressive strength, according to ASTM (2012)
[g]According to CEN (NP 2002)
[h]Bricks loaded on the lateral face (load parallel to the length of the brick)
[i]Aspect ratio correction factors suggested by Heathcote and Jankulovski (1992) for soilcrete blocks were applied in the calculation of the compressive strength
[j]Deformations measured using a Digital Image Correlation technique
[k]Measurement of the relative displacement of the testing platens/measurement conducted directly on the test specimens

bricks (6%). In some cases, different types of specimens are used in the same study, for comparison of results. It is also important to note that, even for specimens with the same shape, dimensions adopted by different authors tend to vary.

Cubic specimens, which are recommended in some standards (Table 1), are the most used. In fact, cubes are relatively simple to prepare and with these specimens it is possible to apply the load in the same direction as would occur in the wall. Some authors, on the other hand, opt to test prisms and/or cylinders with higher aspect ratios, in order to reduce the confinement effect—even though, in this case, the load must be applied perpendicular to the direction of loading in the wall. In these cases, different aspect ratios are adopted, although an aspect ratio of approximately 2 tends to be preferred, since, according to studies on concrete, for an aspect ratio of 2, the failure stress is closer to the unconfined compressive strength (Domone 1994). The use of bricks as test specimens, which is recommended in some standards (Table 1), is also common. Bricks, however, tend to have low aspect ratios, and the confinement effect can lead to unrealistically high estimations of the compressive strength (Aubert et al. 2013).

3.2.2 Specimens Preparation

Cubes and prisms are usually cut from bricks using an appropriate saw, and cylinders are extracted using a rotary corer. Some authors mention difficulties in this process (Fratini et al. 2011; Liberotti et al. 2016), because there may be a tendency for the adobe material to break, especially for adobes from older constructions. Due to this, concern about a possible weakening of the material has been expressed (Liberotti et al. 2016). In some works, it is also mentioned that the top and bottom surfaces of the specimens are smoothed by abrasion, in order to allow a more uniform distribution of the load (Illampas et al. 2014; Liberotti et al. 2016). In most studies, the specimens are air dried and the moisture content is not controlled. In two studies, however, specimens were oven dried before testing (Illampas et al. 2014, 2017). Finally, in some cases, in order to improve the contact between the specimens and the test platens, a regularization element—such as fine sand—is placed in this interface (Quagliarini and Lenci 2010; Silveira et al. 2013).

3.2.3 Testing Procedures

Some authors have carried out load-controlled and others displacement-controlled tests. However, displacement-controlled tests tend to be preferred, because, in this way, the stress-strain curve can also be recorded in the post-peak phase (Fratini et al. 2011). Testing rates vary from 0.1 to 6 MPa/min, in load-controlled tests, and from 0.5 to 4.5 mm/min, in displacement-controlled tests. A wide range of testing rates can thus be found in existing literature. Loading is usually monotonic. However, in one work (Parisi et al. 2015), the loading procedure included two initial cycles in the elastic range in order to improve the contact between the specimen and the platens.

In general, standards recommend testing the specimens in the same direction as would occur in the wall. For cubic and brick specimens, it is possible to accomplish this, and this is the direction generally adopted. However, due to the typical low height of adobe bricks, prisms and cylinders with aspect ratios greater than one tend to be cut horizontally along the length or width of the bricks and thus are tested perpendicular to the loading direction in the wall.

The issue of the influence of the loading direction was recently addressed by Rodríguez-Mariscal et al. (2018). In this study, cubic specimens were tested in two different directions: longitudinal and transversal (relative to the loading direction in the wall, i.e. to the moulding direction). The results obtained pointed to some anisotropy of the material, thus suggesting that the loading direction may, in fact, be a relevant factor that should be considered.

3.2.4 Deformation Measurement

Measurement of the deformation suffered by the specimens during the test is important to assess the behaviour of the material. In order to do this, authors have opted for two different strategies: (i) to measure the displacement in the testing machine (in most cases, the relative displacement of the testing platens), or (ii) to measure the deformation directly on the test specimens. The first method is more commonly adopted, because measurement is easier to conduct in this way. The second option is usually carried out using displacement transducers, which measure the axial displacement between two points of the specimen (Silveira et al. 2013; Rodríguez-Mariscal et al. 2018). This measurement can be difficult to carry out, due to the difficulty in fixing the displacement transducers to the test specimens, since the adobe material tends to be very fragile and due to the small dimensions of some specimens (Illampas et al. 2014; Parisi et al. 2015). The Digital Image Correlation technique has also been used for this purpose (Aguilar et al. 2017).

Comparisons of the results obtained with the two methods have been carried out, and it was observed that the two methods lead to very different results (Silveira et al. 2013; Aguilar et al. 2017; Rodríguez-Mariscal et al. 2018). The second method provides more accurate results (Silveira et al. 2013; Rodríguez-Mariscal et al. 2018), because, in this way, only the deformation of the specimen is measured. In the first method, additional deformation is considered—including the deformation suffered by the testing instruments and, in particular, the displacement due to the settlement of the platens on the specimen.

3.2.5 Standards

In many of the studies analysed, the authors do not indicate which standards were followed in the preparation of specimens and conduction of tests. This is particularly true for older studies. In more recent studies, there seems to be a greater concern in considering the existing normative indications.

Some authors opt to follow European Standards or ASTM Standards, which are directed to other materials (such as concrete and natural stone) (Vega et al. 2011; Parisi et al. 2015; Rodríguez-Mariscal et al. 2018). Other authors consider the indications of standards directed to earthen construction (as those presented in Table 1) (Wosick et al. 2014; Adorni et al. 2013; Silveira et al. 2013) or consider the indications of both ASTM Standards and earthen construction standards (Aguilar et al. 2017).

In some cases, the authors mention difficulties in following the normative procedures to test adobes taken from existing constructions, since most standards directed to earthen construction seem to be more focused on new construction. Overall, authors point to the need for consensus regarding specimen geometry and testing procedures since different standards tend to provide different indications.

3.3 Compressive Strength

The majority of studies used new bricks (56%), produced specifically for the research conducted, but a significant part of studies involve bricks collected from existing constructions, some hundreds or even thousands of years old. The lowest compressive strength of 0.18 MPa was obtained in a research conducted in Turkey from Liberotti et al. (2016). This study tested adobe cubes collected from residential and monumental buildings from the beginning of the fourth millennium to the end of the third millennium B.C. The small value obtained is assigned to a faulty production process. Other authors (Fratini et al. 2011; Adorni et al. 2013; Silveira et al. 2013) present values around 0.20 MPa, all of them with tests conducted on bricks collected from existing constructions. These authors also point that the constructions were built using adobe bricks made without production control, causing important differences, even within the same adobe construction. This may have contributed to bricks with a lower resistance.

Wosick et al. (2014) studied typical adobe bricks from the southwestern United States, having obtained in one of the samples a compressive strength value of 8.89 MPa. However, this is a value that highly stands out of the researches conducted in the rest of world, as most of the studies do not surpass a compressive strength of 3.0 MPa.

There are several factors that can influence the compressive strength of adobe bricks. The level of damage, conservation state and exposition time (Rivera and Muñoz 2005; Fratini et al. 2011; Meli 2005) can have a strong impact in the compressive response. In addition, the composition of the soil, the moisture content and the proportion of natural minerals greatly affect the behaviour of adobe bricks (Wu et al. 2013; Meli 2005). While some authors point out that the geometry size and the aspect ratio of the specimens are also one of the factors for the existence of different results (Illampas et al. 2014; Adorni et al. 2013; Wu et al. 2013;

Rodríguez-Mariscal et al. 2018), others state that the size doesn't alter significantly the results (Wosick et al. 2014). More uniformly, authors agree that the bricks should be placed in the construction of walls in such a way so that the vertical loads are applied along the direction the bricks were manually pressed (Vega et al. 2011; Illampas et al. 2017).

The use of additives in the preparation process of the bricks can increase the compressive strength and the overall behaviour of the adobe bricks (Vega et al. 2011; Adorni et al. 2013; Wu et al. 2013; Illampas et al. 2017). The additives may be natural (animal dung, straw) or artificial (plastic fibres) There seems to be an optimal percentage of additives to be added as they decrease density, witch, after a certain point, causes the decrease of the compressive strength (Wu et al. 2013; Illampas et al. 2017). Moreover, the use of natural additives can cause problems with time, as the additives might degrade (Adorni et al. 2013).

3.4 Stress-Strain Curves

In order to fully understand the mechanical behaviour of adobe bricks, it is desirable to know the complete stress-strain relationship. Several authors propose design constitutive equations obtained in experimental tests (Silveira et al. 2013; Parisi et al. 2015; Aguilar et al. 2017; Rodríguez-Mariscal et al. 2018), to be used in numerical modelling for validation of experimental results.

Figure 1 shows the stress-strain chart obtained in Parisi et al. (2015) for 34 specimens tested, where it is possible to identify the existent different phases. The initial phase corresponds to the contact adjustment between the plate and the specimen (Aguilar et al. 2017; Rodríguez-Mariscal et al. 2018), followed by a linear evolution, the rise until peak strength is reached and the softening behaviour after it, with a logarithmic decay. The specimens response is quite scattered and further research, specially in the post-peak softening range, is required, as pointed out by different researchers (Illampas et al. 2014; Parisi et al. 2015). The elastic-perfectly

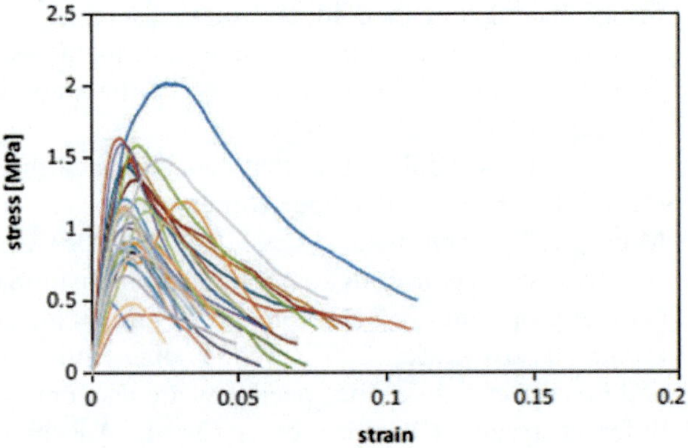

Fig. 1 Experimental stress-strain curves for adobe bricks in compression (from Parisi et al. (2015))

plastic model may be used for simplified structural models for design purposes (Parisi et al. 2015). Aguilar et al. (2017) concluded that the failure mode influences the stress-strain behaviour along with variability of the mixture content and boundary conditions of the specimens tested. All this originates a very high variability in the testing results.

The addition of natural fibre significantly modifies the adobe behaviour, altering the stress-strain relationship (Quagliarini and Lenci 2010; Illampas et al. 2017).

3.5 Young's Modulus

There is a lack of a standard process for the determination of the Young's modulus of earth construction, in general, and of adobe bricks, in particular. This gives rise to the existence of different strain measurements (tangent or secant) and points considered in the computation of the Young's modulus, also referred to as the modulus of elasticity. Stress-strain relationships are non-linear and the points where the measurements are made have a crucial influence in the results. Some authors determined this modulus at 1/3 of the maximum stress, others at 1/2 and others between 1/3 and 2/3 of the maximum compressive strength. For each kind of computation, the ranges of results are still very wide. When calculating the Young's modulus at 1/3 of the peak stress, the results obtained ranged between 145 and 25,000 MPa. When considering a secant between the start and the end of the linear segment of the stress-strain relationship, the results vary between 15 and 667 MPa. Also, a range of 41–289 MPa is obtained for initial tangent and tangent at 60% of the peak stress. Illampas et al. (2014) and Illampas (2017) considered the Young's modulus as the secant between 5 and 50% of the peak stress, which is the stress range over which deformations tend to increase linearly.

The high variation of values translates, primarily, the computation mode or type of measurement adopted, but also the composition of the soil or additives used. Parisi et al. (2015) proposed a linear correlation between compressive Young's modulus and peak compressive strength, for the specimens analysed. However, due to the differences stated, it is not possible to expand the application of empirical relationships, and the Young's modulus must be regarded as an indicator of the stiffness of the brick (Rodríguez-Mariscal et al. 2018).

The composition of the soil plays an important role in the values obtained. Quagliarini and Lenci (2010) found that an increasing coarse sand content in the mixture increases the modulus of elasticity, independently of the straw content. Additionally, the straw content controls the plastic behaviour and influences the failure mode of the adobe bricks. Illampas et al. (2017) concluded that fibers could have an adverse effect on stiffness and should be used with care. Also, Silveira et al. (2013) implies that the addition of a significant fraction of air lime binder is responsible for the high values of Young's modulus obtained (25,000 MPa).

3.6 Strain at Peak Strength

Similarly to other parameters, there is also a great variability on the values of strain at peak strength, which range from 0.024% to higher than 10%. The high deformability verified in Illampas et al. (2014) is assigned to a hastily placement of the adobe mixture in moulds and to a reduced level of compaction during casting of the Cypriot adobes studied. The lower value for strain at peak strength was obtained by Silveira et al. (2013) due to same reason pointed for a high value of Young's modulus: the use of air lime binder as an additive.

Other authors also suggest the influence additives may have in the evolution of strain. Quagliarini and Lenci (2010) conclude that adobe bricks without natural fibres or clay will have a higher peak strength, with a brittle behaviour, i.e., a small strain value.

The use of natural materials and the traditional manual compacting process may cause plastic deformation effects that affects the values of the strain at peak stress (Wu et al. 2013). Illampas et al. (2014) suggests the material deformation capacity should be considered in the compressive behaviour assessment along with the compressive strength and stresses developed.

3.7 Poisson's Ratio

From the analysis of Table 3, it is clear that there is the need for more studies on the Poisson's ratio. From the 16 studies presented, only one effectively calculated this factor for adobe bricks. Again, there is no standard process specified for adobe bricks for the determination of Poisson's ratio.

Silveira et al. (2013) calculated the Poisson's ratio by adapting an expression from ASTM C469/C469M-10 standard (2010), using the ratio between the transverse strain at mid-height of the specimen produced by the stress corresponding to 40% of peak stress and the longitudinal strain produced by the stress corresponding to 40% of peak stress. This resulted in a value of approximately 0.10.

Wu et al. (2013) states that tangent modulus and Poisson's ratio increase with the ratio of stress to peak strength, though more studies are required on this subject.

3.8 Failure Mode

Different authors observed the failure of the specimens tested by splitting with vertical cracks (Illampas et al. 2014; Wu et al. 2013; Reddy and Gupta 2006; Reddy and Vyas 2008; Walker 2004) and, sometimes, bulging was also observed (Illampas et al. 2014). Compressive crushing and large inelastic deformations on the lateral sides was verified in Illampas et al. (2014). Most of the authors agree the mode of failure is influenced by the loading direction (Illampas 2017; Rodríguez-Mariscal et al. 2018).

The behaviour of adobe brick is highly dependent on the relation between soil, additives and clay content. The introduction of natural fibres has shown to have an effect on the failure mode of adobe bricks (Illampas et al. 2014; Quagliarini and Lenci 2010). Fibers allow a redistribution of internal forces, delaying failure and providing a more ductile behaviour (Illampas et al. 2014; Quagliarini and Lenci 2010).

3.9 *Comparison with Normative Limits*

Different norms have different limits. The specimens' dimensions are distinct in each standard, as well as the test rate or special specifications for the conduction of the tests. This means that the limits stated can only be applied if the conditions referred were observed.

The specimens' dimensions are one of the criteria standards use, with specific dimensions or ratio between them to comply with. Some authors used cubes and others used prims, which by itself limits the use of certain norms. In addition, some of the norms specify a particular test rate to be used in the compression tests and most authors do not indicate the test rate used. Finally, the strength limit is set either considering the minimum admissible mean strength or the minimum strength for an individual unit, or for a specified number of best values obtained in a set of specimens.

The studies analysed in this chapter were examined in terms of the applicability of the norms stated in Table 1, its limits were verified, and the results are summarized in Table 4.

The New Zealand Standard (NZS 1998) is the most applicable norm, due to the most general values accepted for the aspect ratio of specimens, including cubes or prisms. Under this standard, from the 16 studies analysed, 8 verify the strength limit. For the 14.7.4 NMAC (2015), from the 5 studies where the norm could be

Table 4 Application and verification of norms

	Possible applications	Verification of strength limit
14.7.4 NMAC (RLD 2015)	5	4
Norma E.080 (MVCS 2017)	8	5
NZS 4298 a (SNZ 1998)	16	8
Pima County Standard (PCDS 2012)	14	6
Australian Handbook (Walker 2002)	14	-

applied, 4 verify the strength limit. 5 experimental programs from the 8 papers where Norma E.080 (2017) could be apply also verify the strength limit. Also, Pima County Standard (Standard for Earthen IRC Structures 2012) could be applied in 14 of the studies evaluated, with 6 of them complying with the strength limit stated.

In a total of 16 studies considered, 69% verify one of the norms. The ones that do not verify any norm correspond to adobe specimens collected from existing constructions that exhibit a very low compressive strength.

4 Conclusions and Final Remarks

In order to understand the behaviour of adobe masonry, it is fundamental to comprehend the mechanical characteristics of adobe bricks. Standards and normative documents with clear indications for the mechanical testing on this subject are still very scarce and not very complete. For this reason, researchers from different parts of the world use different test procedures and distinct types of specimens, obtaining, consequently, results that cannot be directly comparable. In this chapter, a review of the results regarding mechanical characterization of traditional adobe bricks obtained by different authors is presented, specifically when the bricks are under simple compression. In addition, a compilation of the technical recommendations for the mechanical testing of adobe bricks was gathered from different normative documents.

This chapter explores the compressive strength, stress-strain relationships, Young's modulus and Poisson's ratio values obtained under compressive testing of adobe bricks. The results obtained were compared to what is stated in the existing normative documents and it was found that 69% of the studies analysed verify at least one of the norms.

Regardless of the specimen type tested, considerable deviations were noted in the mechanical properties of adobes reported in this chapter. These deviations are primarily attributed to the inherent heterogeneity of the material itself, arising from the adoption of empirical production methods and the lack of standardised criteria for the selection of raw materials. Nevertheless, they may also be explained by the adoption of different testing procedures. This renders the need for standardised procedures for the mechanical testing of adobes imminent.

For adobes from an existent construction, the level of damage, conservation state and exposition time can have an important impact on the compressive behaviour (Rivera and Muñoz 2005; Fratini et al. 2011; Meli 2005). The use of additives can improve the overall behaviour of adobe bricks but research suggests that this will happen up until a certain amount of additives (Wu et al. 2013; Illampas et al. 2017).

Several stress-strain relationships are proposed by different authors, with the indication that the elastic-perfectly plastic model may be used for simplified structural models for design purposes (Parisi et al. 2015). The determination of the Young's modulus also has different perspectives and, thus, different results. Authors use different strain measurements (tangent or secant) and different points considered in the computation. From all the parameters analysed, the Poisson's ratio is the one least studied, requiring more research on the theme. Again, there is a lack of a standard process for the determination of Poisson's ratio.

The material itself presents high variability. Furthermore, there is an inexistence of standardized methods for testing and for measuring the different mechanic parameters. All this combined makes it impossible to directly compare results of different authors and infer widely applicable conclusions. More research is needed on this subject focusing on finding common ground between studies all over the world. By understanding the behaviour of adobe bricks, adequate procedures for the protection and rehabilitation of existent construction may be suitably prescribed. The development of technical standards with uniform procedures, allowing the adequate characterization of the mechanical properties of adobe bricks and earthen materials in general, is vital for an adequate repair and strengthening of structures with this material, which sometimes constitute important built heritage.

References

Adorni E, Coïsson E, Ferretti D (2013) In situ characterization of archaeological adobe bricks. Constr Build Mater 40:1–9

Aguilar R, Montesinos M, Uceda S (2017) Mechanical characterization of the structural components of Pre-Columbian earthen monuments: Analysis of bricks and mortar from Huaca de la Luna in Perú. Case Stud Constr Mater 6:16–28

Aubert JE, Fabbri A, Morel JC, Maillard P (2013) An earth block with a compressive strength higher than 45 MPa! Constr Build Mater 47:366–369

Blondet M, Vargas J (1978) Investigación sobre vivienda rural. Pontifical Catholic University of Peru (In Spanish)

ASTM C42/C42M–12 (2012) Standard test method for obtaining and testing drilled cores and sawed beams of concrete. West Conshohocken, ASTM International

ASTM C469/C469M–10 (2010) Standard test method for static modulus of elasticity and Poisson's ratio of concrete in compression. ASTM International, West Conshohocken

Domone PLJ (1994) Strength and failure of concrete. In: Illston JM (ed) Construction materials: their nature and behaviour. E&FN Spon, London, UK, pp 155–168

Fratini F, Pecchioni E, Rovero L, Tonietti U (2011) The earth in the architecture of the historical centre of Lamezia Terme (Italy): characterization for restoration. App Clay Sci 53(3):509–516

Heathcote K, Jankulovski E (1992) Aspect ratio correction factors for soilcrete blocks. Aust Civil Eng Trans Inst Eng Aust CE34(4):309–312

Illampas R (2013) Experimental and computational investigation of the structural response of adobe structures. PhD Thesis, University of Cyprus, Nicosia, Cyprus

Illampas R, Ioannou I, Charmpis DC (2014) Adobe bricks under compression: experimental Investigation and derivation of stress-strain equation. Constr Build Mater 53:83–90

Illampas R, Loizou VG, Ioannou I (2017) Effect of straw fiber reinforcement on the mechanical properties of adobe bricks. In: Proceedings of the 6th Biot conference on poromechanics, Paris, France

Liberatore D, Spera G, Mucciarelli M, Gallipoli MR, Santarsiero D, Tancredi C (2006) Typological and experimental investigation on the adobe buildings of Aliano (Basilicata, Italy). In: Lourenço PB, Roca P, Modena C, Agrawal S (eds) Proceedings of the 5th international conference on structural analysis of historical constructions, Macmillan India, New Delhi, India, pp 851–858

Liberotti G, Rovero L, Stipo G, Tonietti U (2016) Mechanical investigation on adobe samples belonging to the archaeological site of Arslantepe (Malatya, Turkey). J Mater Environ Sci 7(10):3656–3666

Meli R (2005) Experiencias en México sobre reducción de vulnerabilidad sísmica de construcciones de adobe. In: Proceedings of the SismoAdobe2005: international seminar on architecture, construction and conservation of earthen buildings in seismic areas (CD-ROM), Pontifical Catholic University of Peru, Lima, Peru

New Mexico Code. 14.7.4 (2015) New Mexico earthen building materials code. New Mexico Administrative Code, Construction Industries Division of the Regulation and Licensing Department (RLD), New Mexico

Norma E.080 (2017) Diseño y construcción con tierra reforzada, Ministerio de Vivienda, Construcción y Saneamiento (MVCS), Lima (In Spanish)

NP EN 1052–1 (2002) Methods of test for masonry—Part 1: determination of compressive strength. European Committee for Standardization (CEN), Brussels. Instituto Português da Qualidade (IPQ), Caparica (In Portuguese)

NZS 4298 (1998) Materials and workmanship for earth buildings, Standards New Zealand (SNZ), Wellington

Parisi F, Asprone D, Fenu, Prota A (2015) Experimental characterization of Italian composite adobe bricks reinforced with straw fibers. Compos Struct 122:300–307

Quagliarini E, Lenci S (2010) The influence of natural stabilizers and natural fibres on the mechanical properties of ancient Roman adobe bricks. J Cult Herit 11(3):309–314

Reddy VBV, Gupta A (2006) Strength and elastic properties of stabilized mud block masonry using cement-soil mortars. J Mater Civ Eng 16(5):472–476

Reddy VBV, Vyas UChV (2008) Influence of shear bond strength on compressive strength and stress–strain characteristics of masonry. Mater Struct 41:1697–1712

Rivera J, Muñoz E (2005) Caracterización estructural de materiales de sistemas constructivos en tierra: el adobe. Revista Internacional de Desastres Naturales, Accidentes e Infraestructura Civil 5(2):135–148 (In Spanish)

Rodríguez-Mariscal JD, Solís M, Cifuentes H (2018) Methodological issues for the mechanical characterization of unfired earth bricks. Constr Build Mater 175:804–814

Silveira D, Varum H, Costa A (2013) Influence of the testing procedures in the mechanical characterization of adobe bricks. Constr Build Mater 40:719–728

Standard for Earthen IRC Structures (2012) Pima County Development Services (PCDS), Tucson

Vega P, Juan A, Ignacio M, Morán J, Aguado P, Llamas B (2011) Mechanical characterisation of traditional adobes from the north of Spain. Constr Build Mater 25:3020–3023

Walker P (2002) The Australian earth building handbook, HB 195-2002. Standards Australia, Sydney

Walker P (2004) Strength and erosion characteristics of earth blocks and earth block masonry. J Mater Civ Eng 16(5):497–506

Wosick E, Gebremariam T, Weldon B, Bandini P, Al-aqtah U (2014) Strength characteristics of typical adobe material in the southwestern United States. In: Proceedings of the 9th international masonry conference, Guimaraes, Portugal, July 2014

Wu F, Li G, Li H-N, Jia J-Q (2013) Strength and stress-strain characteristics of traditional adobe block and masonry. Mater Struct 46(9):1449–1457

Mechanical Characterization of Adobe Masonry

Cristina Oliveira, Dora Silveira, Humberto Varum, Fulvio Parisi, Lorenzo Miccoli, Mario Solís, José D. Rodríguez-Mariscal, and Nicola Tarque

Abstract The characterization of the mechanical properties and behaviour of adobe masonry is fundamental for the understanding of the structural behaviour of adobe constructions. Thus, in the last decades, experimental studies focused on this topic have been carried out by different authors. Many of the existing experimental works, however, were carried out aiming to support broader studies focused on the seismic behaviour of adobe constructions and are not very detailed. Moreover, authors tend to adopt different procedures in their experimental work, since there are few indications in existing standards for testing adobe masonry. The wide variety in materials used, both for the adobes and mortars, further complicates this work, making it difficult to compare results obtained in different studies. This chapter provides an overview of the indications given by standards and other technical recommendations for the mechanical testing of adobe masonry. It presents a review of existing research on the mechanical behaviour of adobe masonry, addressing studies that focus on: (i) compression behaviour, (ii) shear behaviour, (iii) joint shear behaviour. It provides a global analysis of the existing knowledge, suggesting improvements for normative documents and identifying future research needs.

C. Oliveira (✉)
CONSTRUCT-LESE, University of Porto Technology School of Barreiro, Polytechnic Institute of Setubal, 2839-001 Lavradio, Portugal
e-mail: cristina.oliveira@estbarreiro.pt

D. Silveira
ADAI-LAETA, Itecons—Instituto de Investigação e Desenvolvimento Tecnológico para a Construção, Energia, Ambiente e Sustentabilidade, 3030-289 Coimbra, Portugal
e-mail: dora.silveira@itecons.uc.pt

H. Varum
CONSTRUCT-LESE, Civil Engineering Department, Faculty of Engineering of the University of Porto, 4200-465 Porto, Portugal
e-mail: hvarum@fe.up.pt

F. Parisi
Department of Structures for Engineering and Architecture, University of Naples Federico II, Via Claudio 21, 80125 Naples, Italy
e-mail: fulvio.parisi@unina.it

H. Varum et al. (eds.), *Structural Characterization and Seismic Retrofitting of Adobe Constructions*, Building Pathology and Rehabilitation 20,
https://doi.org/10.1007/978-3-030-74737-4_4

Keywords Adobe masonry · Mechanical behaviour · Experimental tests · Standards

1 Introduction

Adobe construction is used throughout the world, in different regions with different traditions. It has been more associated with low-income societies that tend to use the natural resources more readily available and with a lesser cost, though new trends of a sustainable construction are boosting the industry and inciting more research on the theme. Also, as this type of construction has several fragilities—such as poor performance against horizontal loads, like those induced by earthquakes, if not adequately designed and strengthened—more knowledge is required on the mechanical properties and behaviour of adobe masonry, to effectively repair and strengthen the existing constructions.

This chapter presents a review of existing research focused on the mechanical characterization of adobe masonry. It addresses studies that focus on the compressive, shear and joint shear behaviour of this type of masonry.

The materials and specimen geometries tested, the testing procedures, the test results and failure modes, among others, obtained by different authors, are displayed and analysed.

2 Standards and Technical Recommendations

In Table 1, a compilation of standards and technical recommendations for the mechanical testing of adobe masonry specimens is shown. In particular, the following documents were considered: Norma E. 080 from Peru (2017), NZS 4298

L. Miccoli
Division Building Materials, Bundesanstalt für Materialforschung und—prüfung (BAM), unter den Eichen 87, 12205 Berlin, Germany
e-mail: lorenzo.miccoli@bam.de

M. Solís · J. D. Rodríguez-Mariscal
Department of Continuous Media Mechanics and Theory of Structures, University of Seville, San Fernando, 4, 41004 Seville, Spain
e-mail: msolis@us.es

J. D. Rodríguez-Mariscal
e-mail: josedrm@us.es

N. Tarque
GERDIS Research Group, Civil Engineering Division, Department of Engineering, Pontificia Universidad Católica del Perú, Av. Universitaria 1801, San Miguel, Lima, Peru
e-mail: sntarque@pucp.edu.pe

Table 1 Standards and technical recommendations for the mechanical testing of adobe specimens

	Document	Specimens	Test rate	Other indications	Strength limit
Compression test	Norma E. 080 (2017)	• Prisms with slenderness of approximately 3			• Mean strength of 4 best specimens (out of 6) > 0.60 MPa
	NZS 4298[a] (1998)	• Prisms with an even number of bricks and slenderness between 3 and 5		• The tests shall be as per Appendix 2B of NZS 4210 (2001)	
	Australian Handbook (Walker 2002)	• Prisms with slenderness between 2 and 5, but not less than 3 courses high	0.5–1 0 MPa/ min	• Use of 4 mm thick plywood at the top and bottom of specimens • Unconfined strength obtained by applying an aspect ratio correction factor to the measured values	
	DIN 18945 (2013)	• Prisms with slenderness > 1		• Clay bricks with nominal heights < 71 mm must be halved with a saw, then the halves are to be walled together. For the joint, cement mortar has to be used. The joint should be as thin as possible and not thicker than 5 mm. As an alternative to cement mortar, gypsum may be used. For all other formats, the compressive strength of the whole block must be tested	Mean strength of at least 6 samples
Diagonal comp.	Norma E.080 (2017)	• Walls with dimensions of 0.65 × 0.65 m^2 (approx.)			• Mean strength of 4 best samples (out of 6) > 0.025 MPa

[a]Indications for 'standard grade earth construction'

from New Zealand (1998), The Australian Earth Building Handbook from Australia (Walker 2002) and DIN 18945 from Germany (2013).

It is important to note that there are still few standards with indications for the mechanical testing of adobe masonry. Moreover, in general, the existing indications are not very detailed and can vary significantly, from standard to standard.

3 Research on the Mechanical Behaviour of Adobe Masonry

3.1 Compressive Behaviour

Different authors have been conducting experimental tests to understand the compressive behaviour of adobe masonry. Table 2 presents the tests performed by different authors and the main results obtained, including the compressive strength, modulus of elasticity, strain at peak stress and Poisson's ratio.

In the following subsections, some of these works are presented in more detail, organized according to the country in which they were carried out.

3.1.1 Adobe Masonry Specimens: Materials and Geometry

Peru A total of 120 adobe prisms were built and tested by Blondet and Vargas-Neumann (1978) and Vargas-Neumann and Ottazzi (1981). The slenderness ratio (thickness: height) of the specimens varied (1:1, 1:1.5, 1:2, 1:3, 1:4 and 1:5). The adobe bricks used had dimensions of 200 × 400 × 80 mm^3 and were laid on top of each other with mortar in between: 89 specimens were built with mud mortar and 31 with a mortar that was a combination of cement, gypsum and mud. In this chapter, only the results obtained for the adobe prisms built with mud mortar are reported.

It is also important to note to correct the irregularity of the top face of each prism by adding a cement/sand mortar and, thus, obtaining a horizontal surface for testing. Then, two steel plates of 200 × 400 × 20 mm^3 were placed at both ends of each pile and then loaded axially. During the specimen's fabrication some 0.01 m steel bars were left inside each pile (Blondet and Vargas-Neumann 1978; Vargas-Neumann and Ottazzi 1981).

Italy In the framework of a national research project, the mechanical behaviour of Italian adobe masonry constructions was experimentally investigated at multiple scales, from constituents—i.e. adobe bricks and mud mortar (Parisi et al. 2013)—to load-bearing masonry walls with openings (Asprone et al. 2016).

Adobe masonry was composed of adobe bricks reinforced with straw fibres and mud mortar, according to local manufacturing procedures of Sardinia Region, Italy.

Table 2 Simple compression tests on adobe masonry specimens performed by different authors

Reference			Meli (2005)	San Bartolomé and Pehovaz (2005)	Torrealva and Acero (2005)	Yamin et al. (2007)	Wu et al. (2013)	Silveira et al. (2015)	Illampas et al. (2017)	Miccoli et al. (2015)	Müller et al. (2017)	NZS 4210 (2001)
Location			Mexico	Peru	Peru	Colombia	China	Portugal	Cyprus	Germany	Germany	Spain
Specimens and procedures	No. of specimens		Not indicated	4	5	15	9	5	13	5	40	9
	Dimensions (m)[a]	*H*:	Not indicated	0.58	0.43	Not indicated	0.53	1.26	0.28	0.5	0.62	0.58
		W:		0.38	0.25		0.29	1.26	0.45	0.5	0.5	0.5
		t:		0.24	0.25		0.2	0.36	0.3	0.12	0.18	0.16
	Aspect ratio (*H*/*t*)		–	2.4	1.7	–	2.7	3.5	0.93	4.3	3.5	3.75
	Aspect ratio factor ('k_a')		–	0.83	0.77	–	0.84	0.9	0.68	0.96	0.91	0.91
	Adobes	Composition	"Mud"	Soil, coarse sand and straw, in proportion 5:1:1	Soil, coarse sand and straw, in proportion 5:1:1	Clayey soil, with or without natural fibres	Soil with 44% clay-silt and 56% sand, and straw (0.5%)	Sandy soil and air-lime binder	Lean clay (65%) and straw fibres (35%)	Soil with 43% gravel and sand, 45% silt and 12% clay	Not indicated	99% soil (60% sand, 20% silt and 20% clay) and 1% wheat straw (5 cm long fibers)
		Condition	New (produced for the study)	New (produced for the study)	New (produced for the study)	Collected from existing constructions	New (produced for the study)	Collected from existing constructions	New (sampled from local manufacturer)	New (produced for the study)	New (produced for the study)	New (produced for the study)
		Comp. strength (MPa)	≈0.51–1.57	2.94	Not indicated	2.84	1.66	0.47	1.28	5.21	3.58	1.13 (CoV 17.3%)
		Tensile strength (MPa)	≈0.20–0.43[c]	Not indicated	Not indicated	0.49[c]	Not indicated	0.14[d]	0.43[c]	0.52[f]	Not indicated	0.18 (CoV 11.65%)

(continued)

Table 2 (continued)

Reference			Meli (2005)	San Bartolomé and Pehovaz (2005)	Torrealva and Acero (2005)	Yamin et al. (2007)	Wu et al. (2013)	Silveira et al. (2015)	Illampas et al. (2017)	Miccoli et al. (2015)	Müller et al. (2017)	NZS 4210 (2001)
	Mortar	Composition	"Mud"	Soil, coarse sand and straw, in proportion 3:1:1	"Mud"	Clayey soil	Soil with 40–49% clay-silt and 51–60% sand, and straw (0.5%)	Sandy soil and hydrated lime, in proportion 1:3	Soil and straw fibres, obtained from crushed adobes	Soil with 55% gravel and sand, 37% silt and 14% clay	Soil with 55% gravel and sand. 37% silt and 14% clay	Same as adobe
		Comp. strength (MPa)					1.58	0.47	2.06	3.32	3.34[g] 3.78[h]	
		Flexural tensile strength (MPa)						0.26	0.41	1.39	Not indicated	
	Results	Measurement of vertical deformations	Not indicated		Not indicated	Directly on test specimens	Relative displacement of test platens	Directly on test specimens	Relative displacement of test platens	Directly on test specimens	Directly on test specimens	Relative displacement of test platens and sensors attached to specimens
		Comp. strength ('f_c') (MPa)	1.32	0.86	0.85	1.1	0.94	0.33	0.88/1.73[e]	3.28	1.99[g] 2.02[h]	1.44 (CoV 6.11%)
		Unconfined comp. strength ($k_d f_c$) (MPa)	–	0.71	0.65	–	0.79	0.3	0.60/1.18[e]	3.15	1.82[g] 1.84[h]	1.31
		Young's Modulus ('*E*') (MPa)	245		432	98	34	757	15.5/21.8[e]	803	783[g] 899h	460.32 (CoV 24.8%)
		Determination of *E*	Not indicated		Not indicated	Not indicated	Initial tangent	Secant (at 1/3 f_c)	Secant (between 0.25 f_c and 0.50 f_c)	Secant (at 1/3 f_c)	Secant (at 1/3 f_c)	Secant between 1/3 and 2/3 f_c)
		Strain at peak stress (%)					1.7	0.31	8.97/13.35[e]	0.55	0.40[g] 0.45[h]	0.84 (Cov 30.61%)

(continued)

Table 2 (continued)

Reference			Meli (2005)	San Bartolomé and Pehovaz (2005)	Torrealva and Acero (2005)	Yamin et al. (2007)	Wu et al. (2013)	Silveira et al. (2015)	Illampas et al. (2017)	Miccoli et al. (2015)	Müller et al. (2017)	NZS 4210 (2001)
		Poisson's ratio ('v')	–	–	–	–	0.18	0.16	0.3	0.37	–	0.047/0.092/0.114/0.134
		Determination of v	–	–	–	–	At 0.2 f_c	At 0.2 fc	Within the range of 0.25–0.50 f_c	At 1/3 f_c	–	At 0.1/0.2/0.3/0.4 f_c
		Stress versus strain curves?	No	No	No	No	Yes	Yes	Yes	Yes	Yes	Yes

[a] *H*—Height; *W*—Width; *t*—Thickness. [b] Calculated according to NZS 4298 (1998). [c] Hexural tensile strength. [d] Splitting tensile strength. [e] After 3 weeks of curing/ after 43 weeks of curing. [f] Pull-off tensile strength. [g] Mortar M3
[h] Mortar M2

All mechanical tests were performed in the Department of Structures for Engineering and Architecture at the University of Naples Federico II, Italy.

The adobe bricks were 100 × 200 × 400 mm^3 in size and had a mean specific weight equal to 16.80 kN/m^3 with coefficient of variation (CoV) equal to 2%. Sieve analysis provided the following percentages by weight: 26.9% of clay and silt; 70.1% of sand; and 3% of gravel. The soil was thus classified as clayey/silty sand. The internal reinforcement of each brick consisted of randomly oriented straw fibres having a mean diameter of 3 mm (CoV = 7%), mean length of 70 mm (CoV = 34%), and mean percentage by weight of 0.64% (CoV = 21%). To ensure a good level of brick-mortar bond, each brick was handcrafted so as to obtain a grooved surface. The manufacturing process of adobe bricks was almost completely based on the manufacturer's experience. The mortar used for the Sardinian adobe masonry had a composition similar to that of adobe bricks, with the only difference related to the lack of straw fibre reinforcement. The water content of the mortar was equal to 20%.

In Asprone et al. (2016), tests were performed in single-leaf wallets, with a size of 610 × 650 × 150 mm^3. Each specimen was composed of six layers of adobe bricks, alternated with 10 mm-thick mortar joints. No specific rules were adopted for the curing of specimens, in order to reproduce the real conditions of local manufacturing practice. For potential comparisons between different masonry types, the specimen layout and size were exactly those adopted by Augenti and Parisi (2010) in a previous experimental study on the compressive behaviour of tuff stone masonry.

Portugal Five full-scale adobe panels were tested in Silveira et al. (2015) for compression. The wall panels were built with the dimensions of 1260 × 1260 360 mm^3, using two and half adobes horizontally and nine rows of adobe blocks vertically. The adobe blocks used were collected from a building under demolition, made of arenaceous soil with a reduced silt-clay fraction and ail-lime in 25–40% percentage. The size of adobe bricks was 460 × 320 × 120 mm^3, with specific weight of 15 kN/m^3, compressive strength of 0.466 MPa (CoV = 34%), modulus of elasticity of 13.068 MPa (CoV = 32%) and splitting tensile strength of 0.137 MPa (CoV = 65%). The mortar used was produced in the laboratory of Aveiro University, using a ratio of 1:3 for the proportion of hydrated lime and soil, in terms of bulk volume. Its specific weight is of 17 kN/m^3 (CoV = 7%), compressive strength of 0.469 MPa (CoV = 24%), and flexural strength of 0.257 MPa (CoV = 19%), determined in accordance with the indications of EN 1015-11 (1999).

Germany In Miccoli et al. (2015), earthen materials for the experiments were sourced from a local manufacturer of prefabricated earthen building products. Earth blocks were produced by a mechanised moulding procedure (no compression) with a size of 240 × 115 × 71.5 mm^3. Earth blocks had a bulk density of 1809 kg/m^3. Earth mortar was provided as pre-mixed dry mortar and was mixed with water for sample preparation, according to DIN 18946 (2013).

Then, earth block masonry panels, of size 500 × 500 × 115 mm^3, were produced, by laying six earthen unit courses. The courses were connected by 20 mm joints, wider than what the standard recommends, since wider joint widths are often observed in historic earth block masonry. These earth block masonry samples had a bulk density equal to 1870 kg/m^3. Samples were built without pre-wetting the earth blocks. After production, the panels were stored in a climate room at 23 °C and 50% relative humidity, for drying. Samples were removed from the climate room shortly before strength testing took place.

Similar procedures were followed by Muller et al. (2017), concerning the origin of earthen materials and the exterior dimensions of the earth blocks. In this research, hollow earth blocks were produced by a mechanised moulding procedure (no compression), with a bulk density of 1476 kg/m^3. Two types of earth mortars were used, with a bulk density of 1901 kg/m^3 (M2) and 1991 kg/m^3 (M3). Earth mortar was provided as pre-mixed dry mortar and was mixed with water for sample preparation, according to DIN 18946 (2013).

Using these earth blocks and mortars, single-leaf earth block masonry panels with 620 × 500 × 175 mm^3 were produced, by laying six earthen unit courses. Panels were built with pre-wetting of joint-contact surfaces, by submerging them into water of 10 mm height, for approximately 10 s. After production, the panels were stored in a climate room at 23 °C and 50% relative humidity, for drying. The drying process was ended when the difference of the samples weight was less than 0.2%, by weight, within 24 h. Samples were removed from the climate room shortly before the compression tests.

Spain In the studies developed by Rodriguez-Mariscal and Solis (2020), the adobes were made of 60% sand, 20% silt and 20% low plasticity clay. A weight portion of 1% of wheat straw was added to the final mixture (fibres 5 cm long). The dimensions of the adobe blocks were 320 × 160 × 80 mm^3. The mortar had exactly the same composition as the bricks, including the straw.

The dimension of the tested walls is defined by the number of rows of bricks, number of bricks per row and thickness of joints. In this case (Rodríguez-Mariscal and Solís 2020), 6 rows and 1.5 bricks per row were used. The thickness of the joints was approximately 20 mm. The recommendations from the European Standard for testing general masonry in compression were followed (EN 772-1 2011).

3.1.2 Testing Procedures

Peru In the work developed by Blondet and Vargas-Neumann (1978) and Vargas-Neumann and Ottazzi (1981), the axial load was applied perpendicular to the mortar joints, with 2.45 kN increments, up to failure of the specimen. The test was force controlled. The axial deformation was measured with extensometers placed in each adobe prism, in the longitudinal faces, and one or two were also placed at the top of each specimen.

Italy The Italian adobe masonry investigated at the University of Naples Federico II, Italy (Parisi et al. 2013), was tested under uniaxial compression, perpendicularly to mortar bed joints. Quasi-static monotonic tests were carried out with displacement control, using a universal testing machine. That machine has a vertical servo-controlled actuator with maximum stroke equal to ±150 mm and load capacity equal to 3000 kN, in compression, and 2400 kN, in tension.

A total of eight linear variable differential transformers (LVDT) were attached to each specimen. Two vertical LVDT, with stroke and gauge length equal to 20 and 200 mm, respectively, were placed over each longitudinal face of the specimen, to measure deformations in the direction of compressive loading. Those LVDT were located at approximately one-third of specimen length and vertically centred across the mid-section of the specimen, allowing the measurement of average compressive deformations. A horizontal LVDT per longitudinal face was placed in line with the mid-section, in order to measure transverse elongations. Those horizontal LVDT had a stroke of 25 mm and a gauge length approximately equal to one-third of the specimen length. Two horizontal LVDT with gauge length equal to 9 mm were also placed on the transverse faces of the specimen (i.e. parallel to the thickness), allowing transverse elongations and splitting failure to be captured. Previous tests by several researchers have shown that even single-leaf masonry prisms can suffer a transverse splitting failure, which consists of one or more vertical cracks (Augenti and Parisi 2010; McNary and Abrams 1985; Vasconcelos and Lourenço 2009).

Regarding the loading protocol, each specimen was first subjected to a couple of load cycles to get an effective contact between the testing machine and specimen. Afterwards, the uniaxial compressive load was first increased and then decreased down to approximately one-half of the peak strength, according to a monotonically increasing displacement with constant rate equal to 0.01 mm/s. That procedure allowed the authors to measure both the rising and post-peak descending branches of force–displacement diagrams, and hence to obtain the full macroscopic (i.e. average) stress–strain curves for the adobe masonry under study.

Portugal Asteel frame was used to test the walls in Silveira et al. (2015), using a hydraulic actuator (maximum load potency of 300 kN). The tests were displacement controlled at a rate of 0.010 mm/s, according to EN 1052-1 (1998). In each face of the tested walls, three horizontal displacement transducers and three vertical displacement transducers were used.

Germany Uniaxial compression tests were performed by Miccoli et al. (2015) and Muller et al. (2017), according to EN 1052-1 (1998), using a universal testing machine. Prior to the tests, two I-girders were attached to the lower and upper faces of the panels to introduce the compression forces into the samples. Five samples were tested under displacement control. The loading speed was adjusted so that the failure point was reached after 20–30 min. Deformations were monitored by linear variable displacement transducers (LVDT), parallel and perpendicular to the loading direction, on both sides of the panels.

Spain For the compression tests conducted by Rodriguez-Mariscal and Solis (2020), the upper and lower faces of the wall were covered with a gypsum layer, in order to achieve a uniform load distribution on the wall. Moreover, neoprene and wooden layers were also put between the platen of the testing machine and the upper face of the wall. Figure 1 displays the wall and the distribution of sensors placed for the tests. One set of sensors (2 vertical and one horizontal) was attached to each side of the wall. The arrangement was designed considering the recommendations from the European Standard for testing general masonry in compression (EN 772-1 2011).

The masonry was tested with a servo-hydraulic actuator attached to a load frame (500 kN load capacity) and all tests were displacement controlled (1 mm/min). The peak strength was reached between 10 and 12 min.

3.1.3 Compressive Strength

Peru In Blondet and Vargas-Neumann (1978), it was concluded that the variation of slenderness ratio did not considerably affect too much on the compression strength. However, it was recommended to test prisms of slenderness ratio 1:4. It was also observed that the age of the specimens can be an important factor; however, more tests should be carried out to study this matter. As a preliminary conclusion it can be established that the compression strength for prisms of slenderness 1:4 is between 0.80 and 1.20 MPa, depending on the specimen age.

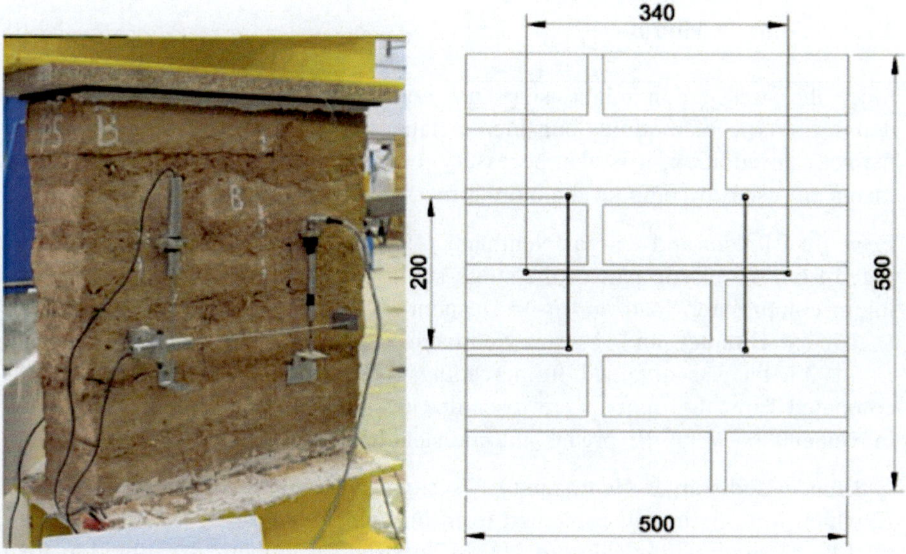

Fig. 1 Arrangement of sensors of compressive tests of walls (Rodríguez-Mariscal and Solís 2020)

Italy The peak compressive strength of the Italian adobe masonry was found to have a mean value equal to 1.32 MPa and CoV = 8%. The mean strength of the whole masonry was higher than that of the straw-reinforced adobe bricks (1.08 MPa). Nonetheless, the mean strength of the masonry fell within the $\mu \pm \sigma$ range of the adobe bricks, given the high dispersion in mechanical properties of straw fibre-reinforced adobe bricks. Such an outcome seems to indicate that the uncertainty level associated with the mechanical behaviour of individual bricks may be reduced at the larger scale of the entire masonry assemblage. From a mechanical viewpoint, this may show that the significant inhomogeneity of single bricks reinforced with straw fibres turns out to be strongly attenuated when considering the whole masonry.

Portugal In Silveira et al. (2015), the compressive strength results of the tested walls varied between 258 and 405 kPa, with a mean value of 331 kPa. The mean compressive strength was found to be lower than the mean compressive strength of the walls' constituent materials. This is due to the fact that the strength of adobe masonry depends on the quality of adherence between adobes and joint mortar, besides on the strength and behaviour of constituent materials.

Germany In Miccoli et al. (2015), the compressive strength values ranged between 2.7 and 3.8 MPa. In Muller et al. (2017), the compressive strength values ranged between 1.99 and 2.02 MPa.

Spain In Rodriguez-Mariscal and Solís (2020), the compressive strength of the walls was 1.44 MPa (CoV of 6.11%).

3.1.4 Young's Modulus

Since the stress-strain relationships are non-linear, it is absolutely necessary to define the type of modulus considered (tangent or secant, and points where it is defined). In addition, it is also necessary to specify which type of strain measurements are used to compute the Young modulus.

Peru In Blondet and Vargas-Neumann (1978), the elasticity modulus was computed from the elastic part of the stress-strain curves, taking the 50% of the maximum compression load and its correspondent deformation. From the specimens tested by Blondet and Vargas-Neumann (1978) a mean elasticity modulus E = 100 MPa was obtained from readings of lateral extensometers. The values computed from the upper were discarded because the readings included relative movements between the prisms and the steal headings.

Blondet and Vargas-Neumann (1978) suggest using values around 170 MPa for E, which were indirectly computed from full adobe wall tests. Later, Tarque et al. (2014a, b) suggested a value of 200 MPa based on numerical calibration for the Peruvian adobe masonry.

Italy The secant Young's modulus of the adobe masonry tested was evaluated at two different levels of compressive stress, namely, one-third ($E_{1/3}$) and one-half ($E_{1/2}$) of peak compressive strength (Parisi et al. 2015). $E_{1/3}$ was found to have $\mu = 432$ MPa and CoV = 32%, whereas the mean value and coefficient of variation of $E_{1/2}$ were $\mu = 425$ MPa and CoV = 25%, respectively.

Portugal The Young's modulus determined in Silveira et al. (2015) ranged between 684 and 821, with a CoV of 8%. Its mean value was of 757 MPa and was determined as the secant modulus of elasticity at one-third of the peak stress, as defined by EN 1052-1 (1998).

Germany In Miccoli et al. (2015) and Muller et al. (2017), the Young's modulus was determined at 1/3 of maximum stress. In the first, it ranged between 587 and 1071 MPa and in the latter it ranged between 783 and 899 MPa.

Spain In Rodriguez-Mariscal and Solis (2020), the tangent modulus between 1/3 and 2/3 of the compressive strength was obtained from the readings of the displacement sensors with the value of 460.32 MPa (CoV of 35.4%). Considering the average global strain obtained from the displacement of the actuator, a value of 218.43 MPa was obtained (CoV of 24.8%).

It was observed that strain measurements from the displacement of the testing machine are erroneous because of the settlement of the platens and the deformation of any capping that is used. These strains overestimate the real strain of the specimen.

3.1.5 Strain at Peak Strength

Peru In Blondet and Vargas-Neumann (1978), the strain measures by upper extensometers were found to be not reliable due to initial relative movement of the steel plate and adobe prisms that can influence the strain readings. All the adobe prisms tested had a constant slenderness ratio 1:4 and were tested after one-month construction.

Italy Regarding the deformation capacity of the adobe masonry in the direction of the compressive loading, the experimental tests performed by Asprone et al. (2016) and Parisi et al. (2015) allowed to characterize the strain at peak compressive strength (ε_p). Such mechanical property was found to have a mean value $\mu = 0.7\%$ and CoV = 19%.

Portugal In Silveira et al. (2015), the ultimate strain was computed at a defined failure point when the compressive stress decreases to approximately 80% of its maximum value. The tests results ranged between 0.40 and 0.62% (CoV of 18%).

Germany In Miccoli et al. (2015), the strain at the ultimate load was in the range of 0.5–1.0%. and in Muller et al. (2017), the same strain was in the range of 0.4–0.55%.

Spain In Rodríguez-Mariscal and Solís (2020), the strain corresponding to the compressive strength was 1.1%.

3.1.6 Poisson's Ratio

The Poisson ratio is computed as the ratio between the horizontal strains measured from the horizontal and vertical sensors at different values of the compressive stress (defined as a ratio of the compressive strength).

Peru In Blondet and Vargas-Neumann (1978), it is suggested to consider a Poisson's ratio between 0.15 and 0.25.

Italy In the case of straw fibre-reinforced adobe masonry (Parisi et al. 2015), the compression tests carried out allowed two estimates of the secant Poisson's ratio to be derived. These were determined at the same levels of stress adopted for the characterization of the secant Young's modulus, that is, at one-third ($\nu_{1/3}$) and one-half ($\nu_{1/2}$) of peak compressive strength. The secant Poisson's ratio at one-third of peak compressive strength ($\nu_{1/3}$) was found to have μ = 0.08 and CoV = 26%. When considering a stress level equal to one-half of peak compressive strength, the secant Poisson's ratio $\nu_{1/2}$ was characterised by μ = 0.17 and CoV = 23%. The latter estimate is more realistic than $\nu_{1/3}$ and is approximately equal to those found after other experimental tests and used in some numerical studies (e.g. Caporale et al. 2015).

Portugal The Poisson's ratio of the walls tested in Silveira et al. (2015) was obtained at 20% of the maximum stress, ranging from 0.04 to 0.29 (CoV = 63%).

Germany In Miccoli et al. (2015), the Poisson's ratio was determined at 1/3 of maximum stress. It ranged between 0.17 and 0.53.

Spain The results obtained in Rodriguez-Mariscal (2020) are listed in Table 3. The results show an increasing of the Poisson ratio with stress. A similar trend was obtained in Wu et al. (2013).

Table 3 Poisson ratio (Rodríguez-Mariscal and Solís 2020)

	Poisson ratio (CoV, %)
0.1 f_{ck}	0.0477 (43.7)
0.2 f_{ck}	0.0918 (24.3)
0.3 f_{ck}	0.1138 (31.4)
0.4 f_{ck}	0.1336 (29.6)

3.1.7 Stress–Strain Curves

Peru Figures 2 and 3 shows some plots of the stress-strain curves obtained from the tests performed in Blondet and Vargas-Neumann (1978). It is clear these tests were force controlled and the inelastic part could not be traced. The load was applied with a hydraulic jack operate manually. The maximum peak strains seem to be larger than the ones for masonry and concrete. More tests, displacement controlled, should be carried out to investigate the inelastic properties of adobe.

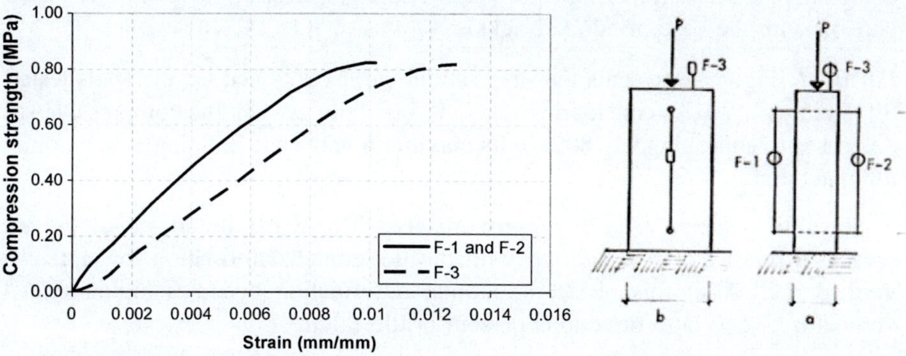

(a) Compression strength vs. strain, specimen C-1 (b) Sketch of the adobe prisms

Fig. 2 Stress-strain curves for axial compression tests on adobe prisms [modified from (Blondet and Vargas-Neumann 1978)]. F-1 and F-2 refers to the lateral extensometers, while F-3 refers to the upper extensometers, **a** Compression strength versus strain, specimen C-1, **b** Sketch of the adobe prisms

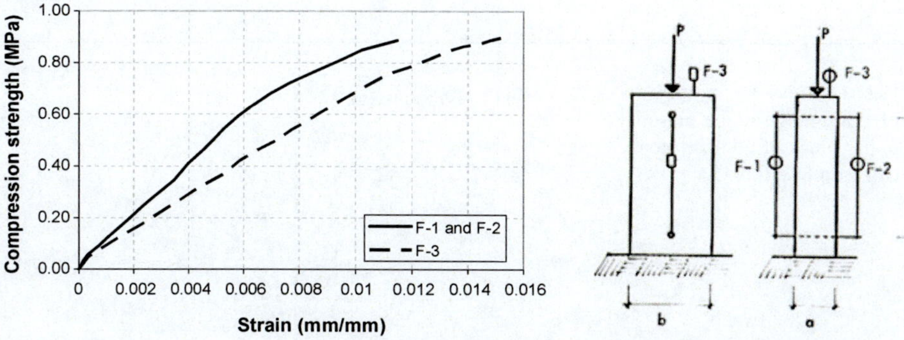

(a) Compression strength vs. strain, specimen C-3 (b) Sketch of the adobe prisms

Fig. 3 Stress-strain curves for axial compression tests on adobe prisms [modified from (Blondet and Vargas-Neumann 1978)]. F-1 and F-2 refers to the lateral extensometers, while F-3 refers to the upper extensometers, **a** Compression strength versus strain, specimen C-3, **b** Sketch of the adobe prisms

Italy In addition to the compressive strain at peak strength (ε_p), the deformation capacity of the adobe masonry was also characterised at a sort of ultimate limit state, which was conventionally assumed to be reached at a 20% strength degradation on the post-peak falling branch of each stress–strain curve. The ultimate compressive strain (ε_u) was found to have a mean value μ = 1.2% and CoV = 24%. If ε_p is assumed to be the limit elastic strain of the masonry, a strain ductility ratio $\mu_s = \varepsilon_u/\varepsilon_p$ can be defined to measure the inelastic deformation capacity of adobe masonry in a simplified manner. This additional mechanical property was found to have μ = 1.73 and CoV = 26% and could be used in elastic-perfectly plastic constitutive models for design purposes (see e.g. Parisi et al. (2015), for a comparison with the case of adobe bricks).

Portugal Figure 4 presents the stress-strain curves obtained for the walls tested in Silveira et al. (2015), considering conventional failure when the compressive stress decreases to approximately 80% of its maximum value and defining at that point the ultimate strain.

Up until a mean value of approximately 35% of maximum stress, the walls present a quasi-linear behaviour, with the horizontal deformation lower than the vertical one. After this phase, horizontal deformation becomes higher than the vertical one, and both directions present brittle failure.

Germany In Miccoli et al. (2015), the stress–strain curves exhibit a short phase of post-peak strain softening under compression, due to its brittle behaviour under uniaxial load, while in Muller et al. (2017), the stress–strain curves exhibit a linear development until approximately 1/3 of the maximum load. Subsequently the curves indicate a non-linear behaviour until the maximum load and finally show a phase of post-peak strain softening. The failure mode under compression loading is rather brittle and the stress–strain behaviour is similar to conventional masonry.

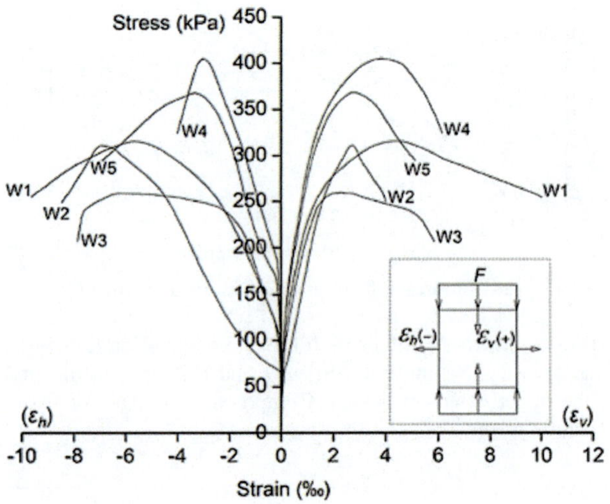

Fig. 4 Stress–strain curves of Portuguese adobe masonry under uniaxial compression (Silveira et al. 2015)

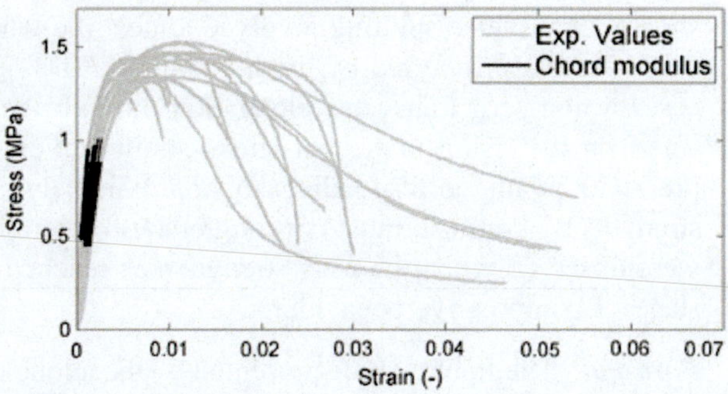

Fig. 5 Stress-strain curve for each wall (Rodríguez-Mariscal and Solís 2020)

Spain Figure 5 shows the curves obtained for the walls tested in Rodriguez-Mariscal and Solis (2020).

3.1.8 Failure Mode

Peru In all cases analysed in Blondet and Vargas-Neumann (1978), the observed failure was brittle, and cracks did not follow a common pattern. For example, in some specimens the cracks started at the central part and went diagonally to the upper part, while in the others the cracks were parallel to the load application (vertical). It should be known that it is hard to observe a non-brittle failure under force control.

Italy In Fig. 6a, b, the crack patterns observed on longitudinal and transverse faces, respectively, of an adobe masonry specimen tested in Parisi et al. (2015) are

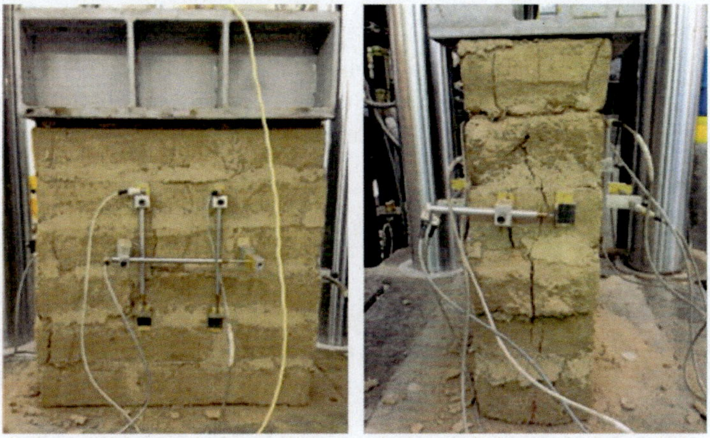

(a) (b)

Fig. 6 Crack patterns of Italian adobe masonry specimens subjected to uniaxial compression: **a** longitudinal face; **b** transverse face (Asprone et al. 2016) shown. Transverse splitting involved almost the whole height of the specimen and was effectively captured by the horizontal LVDT. It could be shown that in most cases a non-zero transverse elongation through the thickness was distinctly measured on the rising branch of stress–strain curves, starting from a vertical (compressive) strain corresponding to approximately one-half of peak compressive strength. By contrast, most crushing cracks on longitudinal faces of each specimen were observed when the peak strength was reached and in the post-peak softening stage of compressive behaviour.

Portugal The failure mode presented by the adobe masonry walls tested in Silveira et al. (2015) was brittle, with failure happening for small deformations after the maximum stress is attained.

The failure pattern shown in the walls consisted mainly of vertical splitting cracks, which mostly appear in the interface adobe bricks and the head joint mortar. Silveira et al. (2015) concluded that head joints may contribute to lower masonry compressive strength as they are potential areas of weakness in the masonry. Moreover, the damage is scattered along the walls' height, at the referred joints.

Minor cracking occurred after loss of the linear behaviour, at around 35% of the maximum stress, though the first visible cracking only occurred ar a mean compressive stress of 62% of the maximum strength.

Germany In Miccoli et al. (2015), failure was usually abrupt after maximum stress was reached, it was visible by vertical or diagonal cracks with sometimes cone shaped cracking pattern, at least one side, sometimes on both sides.

Additionally, in Muller et al. (2017), on the longitudinal side of the samples, failure was visible by vertical cracks above and beneath the butt joints, and mortar spalling in the bed joints. The front side showed a cone-shaped failure pattern. Both are typical failure modes of masonry. The tensile strength of the blocks was exceeded.

Spain In Rodríguez-Mariscal and Solís (2020), the typical failure mode was defined by a vertical cracking pattern. The cracks did not strictly follow the joints of the masonry. Some of the blocks of adobe were also vertically cracked.

3.2 Shear Behaviour

In order to study the shear behaviour of adobe masonry, diagonal compression tests were conducted by different authors and the results related to the compressive strength, tensile strength, shear strength and shear modulus were analysed. Table 4 displays the results obtained in (Meli 2005; San Bartolomé and Pehovaz 2005; Torrealva and Acero 2005; Yamin et al. 2007; Silveira et al. 2015; Miccoli et al. 2015; Rodríguez-Mariscal and Solís 2020).

Table 4 Diagonal compression tests on adobe masonry specimens performed by different authors

Reference			Meli (2005)	San Bartolomé and Pehovaz (2005)	Torrealva and Acero (2005)	Yamin et al. (2007)	Silveira et al. (2015)	Miccoli et al. (2015)	Rodríguez-Mariscal and Solís (2020)
Location			Mexico	Peru	Peru	Colombia	Portugal	Germany	Spain
Specimens	No. of specimens		Not indicated	4	3	10	5	12	5
	Dimensions (m)[a]	*H*:	Not indicated	0.8	0.5	0.75–1.00	1.26	0.5	$H = W = 900$
		t:		0.24	0.25	0.15–0.40	0.36	0.12	160
	Adobes	Composition	"Mud"	Soil, coarse sand and straw, in proportion 5:1:1	Soil, coarse sand and straw, in proportion 5:1:1	Clayey soil, with or without natural fibres	Sandy soil and air-lime binder	Soil with 43% gravel and sand, 45% silt and 12% clay	Sand, silt and clay, in proportion 3:1:1
		Condition	New (produced for the study)	New (produced for the study)	New (produced for the study)	Collected from existing constructions	Collected from existing constructions	New (produced for the study)	New (produced for the study)
		Comp. strength (MPa)	≈ 0.51–1.57	2.94	Not indicated	2.84	0.47	5.21	1.13 (CoV 17.3%)
		Tensile strength (MPa)	≈ 0.20–0.43[b]	Not indicated	Not indicated	0.49[b]	0.14[c]	0.52[d]	0.18c (CoV 11.65%)
	Mortar	Composition	"Mud"	Soil, coarse sand and straw, in proportion 3:1:1	"Mud"	Clayey soil	Sandy soil and hydrated lime, in proportion 1:3	Soil with 55% gravel and sand, 37% silt and 14% clay	Same as adobe
		Comp. strength (MPa)					0.47	3.32	–
		Flexural tensile strength (MPa)					0.26	1.39	–

(continued)

Table 4 (continued)

Reference			Meli (2005)	San Bartolomé and Pehovaz (2005)	Torrealva and Acero (2005)	Yamin et al. (2007)	Silveira et al. (2015)	Miccoli et al. (2015)	Rodríguez-Mariscal and Solís (2020)
Location			Mexico	Peru	Peru	Colombia	Portugal	Germany	Spain
	Results	Shear strength (‘f_v’) (MPa)	0.14	0.11	0.07	0.03	0.03	0.21	0.183
		f_v/f_c (%)	10	13	8	3	8	7	–
		$f_v/(k_a f_c)$ (%)		15	10		9	9	–
		Shear Modulus (‘G’) (MPa)	Not indicated	Not indicated	Not indicated	27	413	397	268/821
		Determination of G				Not indicated	Secant (at 1/3 f_c)	Secant (at 1/3 f_c)	Tangent between 1/3f_c and 2/3 f_c / secant (at 1/3 f_c)

[a]H—Height; W—Width; t—Thickness. [b]Flexural tensile strength. [c]Splitting tensile strength. [d]Pull-off tensile strength

3.2.1 Adobe Masonry Specimens: Materials and Geometry

Peru In Vargas-Neumann and Ottazzi (1981), a first campaign of 10 square wallets of 600 × 600 × 20 mm^3 were built using 200 × 400 × 80 mm^3 adobe bricks, which imply 6 layers of 1½ adobe bricks (Fig. 7). The load was applied at two opposite corners of the wallet. Instrumentation to measure the diagonal deformations were left in each adobe panel, which are used to compute the shear modulus. For the second group of tests (Vargas-Neumann and Ottazzi 1981), 7 panels were built and tested vertically. More precise equipment for load application and to read deformations was used.

Italy The adobe masonry with straw fibre-reinforced bricks investigated at the University of Naples Federico II, Italy (Parisi et al. 2015), was also tested to assess its response to shear loading. The adobe bricks and mortar used for the fabrication of masonry specimens were those previously described in the compression section of this chapter. A mobile experimental set-up was used to avoid any damage to the specimen during handling operations before the test, as shown in Fig. 8. Seven specimens were fabricated according to a running bond scheme and were cured in standard laboratory conditions. Each specimen had a double-leaf resulting in a thickness of 410 mm and consisted of ten masonry layers with 10 mm-thick mortar joints. Therefore, the specimens had squared longitudinal faces with length equal to 1200 mm, according to ASTM-E-519-07 (2007). This is a standard geometry that allows the shear properties of the adobe masonry to be compared with those derived for other masonry types in previous experimental programs [see e.g. Parisi et al. (2013)]. Each specimen was built over a three-layer masonry prism made of hollow concrete blocks and cement mortar joints. The space between the specimen and its support was filled with an ultra-low-density foam material.

In addition, the masonry support was 100 mm-long to leave a free-standing part of the specimen for subsequent application of loading devices at a bottom corner. This allowed a simple application of diagonal compressive loading as described in the following section. It is also noted that the specimens had no passing-through bricks as observed in many existing masonry buildings not designed for earthquake

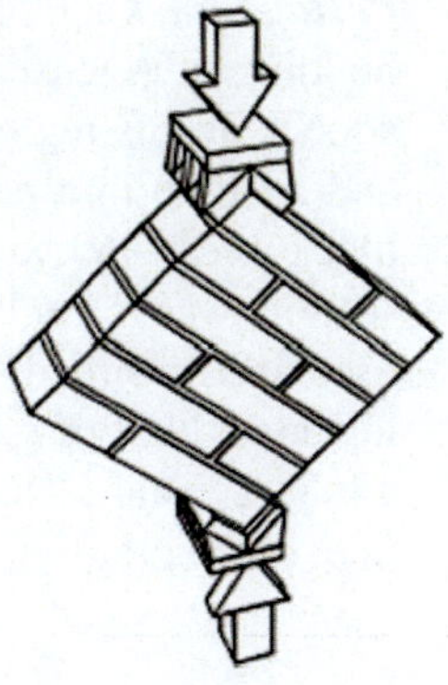

Fig. 7 Scheme of the diagonal compression test [modified from NMX-C-085-ONNCCE (2002)]

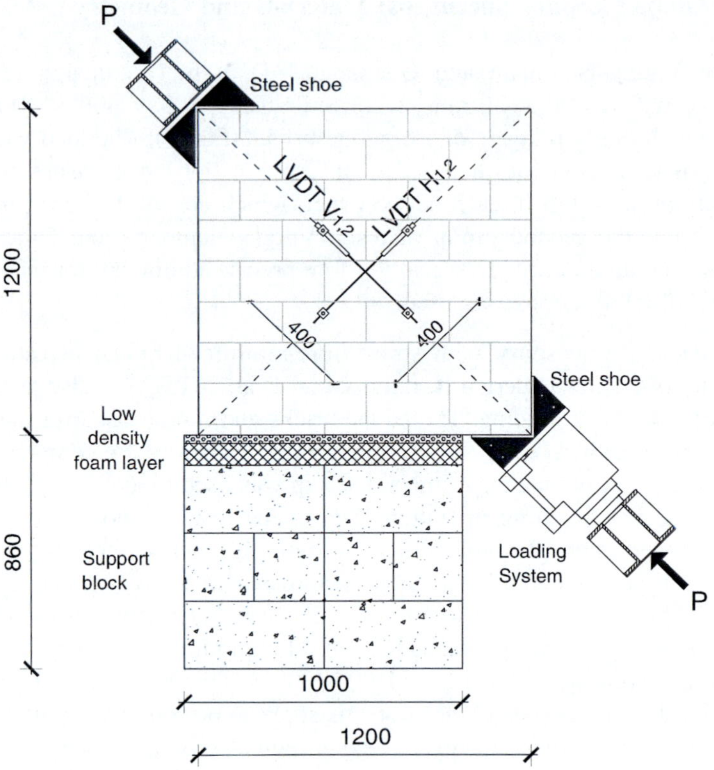

Fig. 8 Scheme of experimental set-up for diagonal compression tests on Italian adobe masonry specimens (Asprone et al. 2016)

resistance. Transverse bricks were just alternated with others at two opposite specimen edges in order to ensure discontinuous head joints. Nonetheless, considering the relatively small length of the specimen, the lack of transverse connections over approximately 1100 mm may be observed even in modern masonry buildings, allowing no influence on the generality of experimental results.

Portugal Five adobe wall panels made with adobes collected from a typical adobe building in Aveiro district under demolition, were used (Silveira et al. 2015). The adobes of this region are made of arenaceous soil with a reduced silt-clay fraction and air-lime in a percentage between 25 and 40%. The dimensions of the adobes blocks were 460 × 320 × 120 mm^3. The mortar, already described in the previous section, was formulated in laboratory and made of a similar composition as the ones used int Aveiro district adobe buildings: hydrated lime: soil ratio 1:3. The soil included 11% of gravel and 4% of clay and silt, and was classified as sand (ISO 14688-2 2004). The tests of the walls took place 60 days after their construction, due to laboratory constraints. It was assumed that the effect of the difference between mortar strength values from the 28th day, as EN 1015-11 (1999) states, until the 60th day was not significant.

Germany Miccoli et al. (2015) also performed diagonal compression tests using the same adobe blocks from the compression tests: earth blocks produced by a mechanised moulding procedure (no compression) with a size of 240 mm × 115 mm × 71.5 mm. The earth mortar was provided as pre-mixed dry mortar and was mixed with water for sample preparation according to DIN 18946 (2013).

In the masonry panels of size 500 mm × 500 mm × 115 mm, the joints were performed with 20 mm thickness, wider than what the standard recommends, due to the fact higher joint widths are often observed in historic earth block masonry. Masonry panels were prepared with wetted blocks, with a bulk density of 1870 kg/ m^3. Past and current practice is actually not to wet earth blocks for building-up earthen masonry. However, earth blocks show generally high-water absorption and suck out moisture from the fresh mortar, which influences the bond between the two panel components greatly. After production, the panels were stored in a climate room at 23 °C and 50% relative humidity for drying. Samples were removed from the climate room shortly before strength testing took place.

Spain In Rodriguez-Mariscal and Solís (2020), the adobes used for the diagonal compression tests had the same dimensions and composition then the ones used for the compression tests: 60% sand, 20% silt, 20% low plasticity clay and 1% wheat straw. The dimensions of the walls were 900 × 900 × 160 mm. The mortar had the same composition as the adobes, including the straw. The thickness of the joints was 20 mm approximately. Each wall was made of 9 rows of adobe, with 2.5 adobes per row.

3.2.2 Testing Procedures

Peru In Blondet and Vargas-Neumann (1978), the 9 wallets were built horizontally and tested in the same position, as seen in Fig. 9a. The maximum load capacity of the hydraulic jack was 100 kN. The load was applied manually at 0.5 kN increments at the wall corners until the specimen failure. Deformations at the diagonals were properly measured; however, some errors were expected due to the movement of the steel heads placed at the wallet corners.

For the second group of tests, in Vargas-Neumann and Ottazzi (1981), the panels were tested vertically as seen in Fig. 9b. The diagonal load (at the panel corners) was applied which a velocity of 2 kN/min.

Italy In Parisi et al. (2015), the diagonal compression tests were carried out through a mobile loading apparatus that included a couple of servo-controlled hydraulic jacks placed at opposite corners of the specimen (Fig. 10). The maximum load capacity of the system was 500 kN both in tension and in compression,

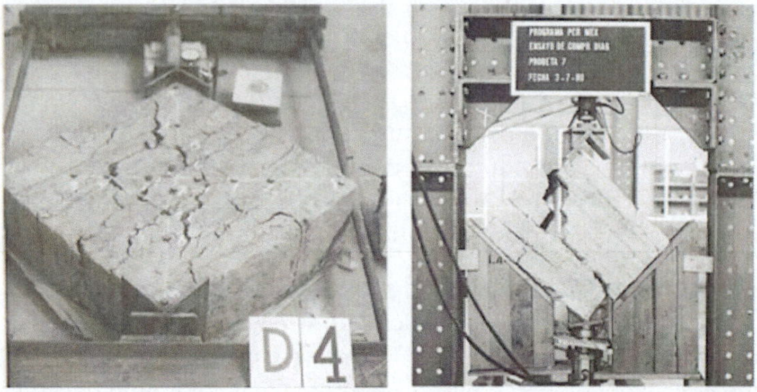

(a) Masonry horizontally tested [16] (b) Masonry vertically tested [17]

Fig. 9 Diagonal compression tests on adobe wallets **a** Masonry horizontally tested (Blondet and Vargas-Neumann 1978), **b** Masonry vertically tested (Vargas-Neumann and Ottazzi 1981)

Fig. 10 Diagonal compression test on adobe masonry specimen (Asprone et al. 2016)

whereas the total stroke of hydraulic jacks was ±75 mm. Rigid steel shoes allowed the application of loading along a diagonal of the specimen, because they are placed between the hydraulic jacks and specimen's corner. Those shoes could distribute the point load over a sufficiently large surface, avoiding a premature crushing failure of the adobe masonry at the loaded corners of the specimen. According to previous experimental programs [see e.g. Parisi et al. (2013)], the space between the specimen and steel shoes was filled with fast-setting and shrinkage-free mortar. The typical failure mode that occurred demonstrates the effectiveness of the test set-up, allowing the shear behaviour of the adobe masonry to be investigated. Rigid steel beams provided contrast to hydraulic jacks and were connected to each other by means of a couple of steel bars.

Two LVDT with a gauge length of 400 mm were installed along diagonals of each longitudinal face of the specimen, particularly in the central zone. This allowed real-time measurement of compressive and tensile deformations induced by diagonal compressive loading.

Each test was performed with displacement control, assuming a constant dis-placement rate equal to 0.01 mm/s. The test was stopped when a complete failure of the specimen was observed and approximately one-half of the peak load was reached on the post-peak descending branch of load–displacement curve.

Portugal For the tests of diagonal compression in Silveira et al. (2015), the walls were wrapped in plastic film in order to control the projection of debris due to the tests. The walls were then placed in a steel frame with rings in corners to easily transport and rotate them. The load was applied through a hydraulic actuator on two steel loading shoes designed according to ASTM E519 (2010). The maximum load potency of the actuator was of 300 kN. The tests were displacement controlled, with the first test conducted at a rate of 0.050 mm/s, while the following ones at 0.025 mm/s in order to comply with the recommendations of ASTM E519 (ASTM-E-519-10 2010). Two vertical and two horizontal rectilinear displacement transducers were placed in each side of the wall.

Germany In Miccoli et al. (2015), diagonal compression tests were performed in compression mode following (ASTM-E-519-10 2010). According to the test, to induce shear forces, panels were turned by 45° around the middle axis with one diagonal of the panels being perpendicular to the loading plates of the testing machine. The stress was introduced by loading plates. The loading plates, placed at the corners of the test samples, were designed according to the standard and failure of the panels never occurred due to excessive compression stress at the corners. Twelve samples were tested, diagonal compression tests were carried out under displacement control. According to the standard test method the loading speed was adjusted in a way that the failure point was reached after 1–2 min.

Spain In Rodríguez-Mariscal and Solís (2020), a load frame with a servo-hydraulic actuator was used for the tests. The RILEM recommendations (1994) and the ASTM Standard (2007) for these tests for masonry were considered for the experimental set-up. The tests were displacement controlled at a rate of 2 mm/min. The maximum load was reached between 1 and 2 min. The load and displacement of the actuator was measured. Moreover, a total of eight displacement sensors were attached to the wall to measure the relative displacements between the points they are attached to (working each sensor as a macrostrain sensor). Two vertical sensors and two horizontal sensors were placed at each side of the wall. A pair of vertical and horizontal strain readings provide a measurement of the shear deformation of the wall. Two different distances (300 and 600 mm) between the fixing points of the sensors were considered for each side of the wall and for each measuring direction.

In these tests, a proper load application and transmission at the corners of the tested wall is a major concern. For this purpose, a specific and novel method was designed. The corners were cut at a 45° angle with a manual radial saw (Fig. 11), so the load was applied at a horizontal surface of 250 mm long, approximately. Thus, the load was transmitted uniformly at more than one row of adobes at the corner, ensuring a proper load transmission from the top to the bottom corner and avoiding the sliding of the top row. In order to achieve a uniform load application and reduce the effect of lack of flatness of the contact surfaces, a layer of gypsum was applied at both corners as a capping (Fig. 12).

Fig. 11 Preparation (cutting) of a corner for a diagonal compression test (Rodríguez-Mariscal and Solís 2020)

Fig. 12 Diagonal compression test (Rodríguez-Mariscal and Solís 2020)

3.2.3 Shear Strength

The ASTM Standard states that the shear stress (τ) is computed following the criteria:

$$\tau_0 = 0.88 \frac{P_{\max}}{A_n} \tag{1}$$

where $P_{\max}$ is the peak compressive load resisted by the specimen, whereas A_n is the gross sectional area of the specimen, that is, $A_n = l\,t$, where l and t are the length and thickness, respectively.

Following the RILEM recommendations, the tensile strength (f_t) of the masonry can be computed from these tests as

$$\tau_0 = 0.707 \frac{P_{\max}}{A_n} \tag{2}$$

Peru Table 5 shows the summarized results obtained from the diagonal compression tests conducted in Vargas-Neumann and Ottazzi (1981), in different wall specimens. The maximum tensile strength, f_t, is evaluated using the following Equation (Brignola et al. 2009).

$$f_t = \frac{1}{\sqrt{2}} \frac{P_{\max}}{l \cdot t} \tag{3}$$

where $P_{\max}$ is the maximum applied load, l is the lateral dimension of the square wallet, and t is the wall thickness.

Table 5 Results of diagonal compression tests on adobe wallets carried out by Vargas-Neumann and Ottazzi (1981)

ID	Maximum tensile stress (MPa)	Shear modulus (MPa)
D-1	0.03	67.0
D-2	0.03	19.4
D-3A	0.033	25.0
D-3B	0.024	15.8
D-4	0.027	17.0
D-5	0.027	90.7
D-6	0.027	11.0
D-7	0.027	90.6
D-8	0.027	17.1
D-9	0.024	24.6
CDPM-1	0.032	50.0
CDPM-2	0.026	34.0
CDPM-3	0.027	40.3
CDPM-4	0.019	—
CDPM-5	0.017	51.8
CDPM-6	0.013	46.7
CDPM-7	0.025	35.8
Mean	0.026	39.8
COV (%)	7.6	61.4

Italy In Parisi et al. (2015), load and displacement recordings collected in the data acquisition system after each diagonal compression test were processed according to the two different procedures of standards (RILEM-TC-76-LUM 1994) and (ASTM-E-519-07 2007). Such standards are based on different mechanical models for the interpretation of diagonal compression tests. Details on the formulation for the estimation of shear stresses and strains at each loading step of the test can be found in the paper by Parisi et al. (2013). It was found that the mean value of τ_0 was μ = 0.094 MPa and μ = 0.076 MPa according to RILEM-TC-76-LUM (1994) and ASTM- E-519-07 (2007), respectively. In both cases, the scattering of experimental data was rather low, as measured by CoV = 6%.

Portugal In Silveira et al. (2015), the mean shear strength obtained in the experimental test was 8% of the mean compressive strength, with a value of 26 kPa and coefficient of variation of 16%.

Germany Miccoli et al. (2015) followed the ASTM Standard (2010) and the shear strength values ranged between 0.15 and 0.30 MPa. The results underlined that the different values of preload applied during the entire drying process of the samples have not relevant influence on the shear strength.

Spain According to this, in Rodriguez-Mariscal and Solís (2020), the obtained average shear strength was 0.183 MPa (CoV of 3.98%). The average value of the shear strain corresponding to the maximum shear stress is 0.0027 (CoV of 42.41%).

The compressive strength of the adobes was 1.13 MPa (CoV = 17.3%), obtained from compression tests of cubic specimens cut from adobe blocks. The tensile strength of the adobes was 0.18 MPa (CoV of 11.65%). It was measured from splitting test of cylindrical specimens. The value of this tensile strength is in very good agreement with that obtained the diagonal compressive tests of adobe masonry.

3.2.4 Shear Modulus

Peru In Blondet and Vargas-Neumann (1978) suggest using a shear module G = 70 MPa (varying from 36 to 90 MPa), and a Poisson module ν = 0.2 (varying from 0.15 to 0.25).

Italy Asecant shear modulus was computed in Parisi et al. (2015) to define a nominal slope for the elastic branch of shear stress–strain diagrams. The secant shear moduli corresponding to one-third of peak shear strength (as defined for the adobe masonry subjected to simple compression) was characterised by mean values approximately equal to or greater than two times the Young's modulus. Therefore, those data were not assumed to be reliable values for the characterisation of the adobe masonry under study. By contrast, the secant shear modulus at one-half of peak shear strength was found to have μ = 290 MPa and CoV = 33%. An alternative estimate of secant shear modulus was also defined. To that aim, a cracking shear strain γ_{cr} was identified on the rising branch of each stress–strain curve, at the attainment of a cracking shear strength nominally defined as $\tau_{cr} = 0.7\tau_0$. This allowed the authors to define a cracking shear modulus $G_{cr} = \tau_{cr}/\gamma_{cr}$, which was found to have μ = 88 MPa and CoV = 13%.

The shear modulus G is evaluated using Eq. (4), considering the 50% of the applied load and the corresponding tangential strain, γ, of the stress-strain curve.

$$G = 0.707\frac{P}{A \cdot \gamma} \tag{4}$$

The tangential strain is given as the sum of the tensile ε_t and compressive ε_c strain [Eq. (5)] evaluated from the relative displacement between two controls points in each wallet diagonal.

$$\gamma = |\varepsilon_t| + |\varepsilon_c| \tag{5}$$

Portugal In Silveira et al. (2015), the shear modulus was determined at one-third of the peak stress, following the equation:

$$G_{sec} = \frac{f_v}{\gamma} \tag{6}$$

The values obtained ranged from 336 to 497 MPa, with a coefficient of variation of 16%. At failure, the secant shear modulus presents a mean value of 10 MPa and at peak stress, 50 MPa.

Germany In Miccoli et al. (2015), the shear modulus was calculated at 1/3 of maximum shear stress. It ranged between 152 and 641 MPa.

Spain According to the ASTM standard, the shear stress–strain behaviour can be obtained from the estimated shear stresses and the measured shear global strains from the displacement sensors.

In Rodríguez-Mariscal and Solís (2020), the shear stiffness modulus was computed as the tangent modulus between 1/3 and 2/3 of the shear strength. The obtained average value is 273 MPa (CoV of 24.36%)

3.2.5 Stress–Strain Curves

Italy Figures 13 and 14 show the shear stress versus shear strain curves of adobe masonry specimens tested at the University of Naples Federico II, Italy (Asprone et al. 2016). In detail, those stress–strain curves are associated with shear stress computation rules according to ASTM-E-519-07 (2007) and RILEM-TC-76-LUM (1994). Except from a single case, all specimens were characterised by a significant strain softening in the post-peak stage of shear behaviour. The processing of such stress-strain curves allowed the shear modulus, peak shear strength at zero confining stress and other mechanical properties to be computed and statistically characterised.

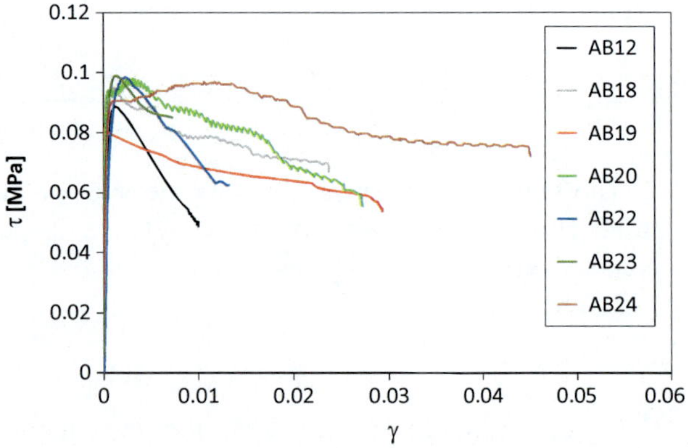

Fig. 13 Stress–strain curves of Italian adobe masonry under diagonal compression according to (Asprone et al. 2016; ASTM-E-519-07 2007)

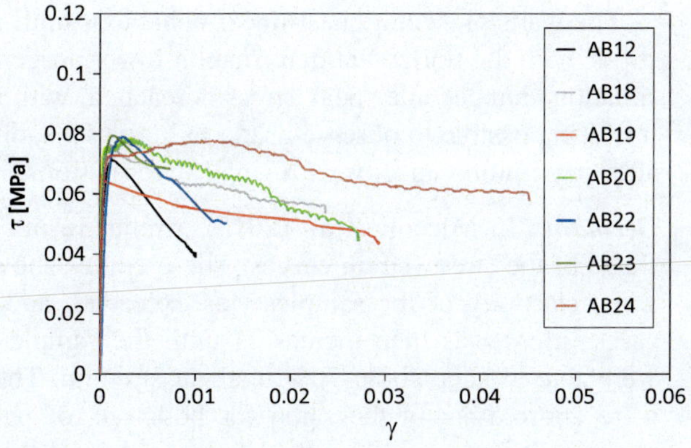

Fig. 14 Stress–strain curves of Italian adobe masonry under diagonal compression according to (Asprone et al. 2016; ASTM-E-519-07 2007)

The stress–strain curves were further processed to define the elastic and ultimate limit states of the adobe masonry in shear. The limit elastic shear strain, denoted as γ_e, was associated with τ_0 and was found to have a mean value $\mu = 0.08\%$ and CoV = 15%. The ultimate shear strain, denoted as γ_u, was associated with the attainment of a 20% strength degradation on the post-peak softening branch. The statistics of γ_u were found to be $\mu = 1.37\%$ and CoV = 52%. Accordingly, a shear strain ductility was also derived, having $\mu = 17$ and CoV = 41%.

Portugal Figure 15 presents the response of the adobe walls tested in Silveira et al. (2015) for diagonal compression regarding stress–strain curves, which are presented to the conventional failure defined when the shear stress decreases to approximately 80% of its maximum value. The shear stress was computed considering only the applied load, as indicated in ASTM E519 (2007).

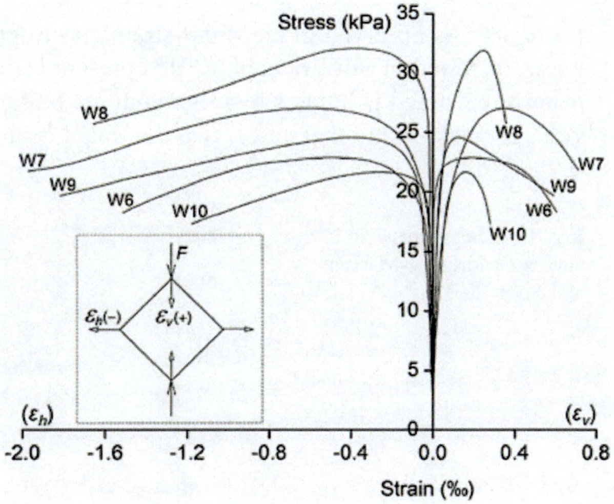

Fig. 15 Shear stress-strain curves (Silveira et al. 2015)

The walls present a quasi-linear behaviour until approximately 75% of the peak stress, with the horizontal deformation lower, in general, than the vertical one. This situation changes after peak stress is reached, with higher horizontal deformations. It is also possible to observe brittle failure, as a sudden reduction of strength occurs after maximum capacity with small deformations.

Germany In Miccoli et al. (2015), similar results were obtained for the graphic shape of the stress–strain curves. These curves showed a distinctive yielding point, when elasticity of the samples was exceeded, and first cracks appeared. In some cases, stress was then increasing until the sample failed. In other cases, samples were able to keep about 70% of shear strength. The maximum deformation values were approximately the same for both sets of panels. This type of stress-strain response is linked to the actual failure mode of the sample.

Spain With the diagonal compression results, stress-strain curves were produced. Figure 16 shows the curves obtained by Rodríguez-Mariscal and Solís (2020). It is possible to identify the yielding point when first cracks appear.

3.2.6 Failure Mode

Italy In Fig. 17, the typical crack pattern of adobe masonry specimens tested in Asprone et al. (2016), is shown. When the peak shear strength was attained, a single diagonal crack was observed approximately along the loading direction. Particularly in the central zone of the specimen, which was not influenced by complex distributions of stresses and strains in the proximity of the loaded corners, the diagonal crack had a stair-stepped shape according to the masonry bond geometry.

Portugal As observed in the stress-strain curves presented in the previous section, the walls tested in Silveira et al. (2015) present brittle failure with a typical damage pattern registered in studies by other authors. Failure happened with a large central vertical crack or, in other cases, two dominant central vertical cracks. As the cracks

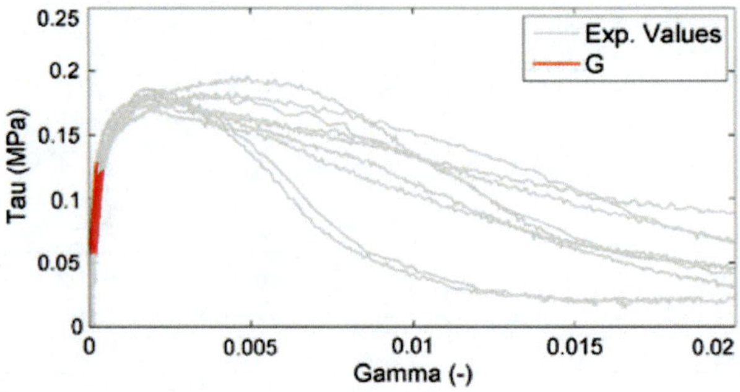

Fig. 16 Shear stress-strain curves (Rodríguez-Mariscal and Solís 2020)

Fig. 17 Typical crack pattern of Italian adobe masonry subjected to diagonal compression (Asprone et al. 2016)

follow along mortar joints in several parts, these regions may be areas of weakness in this type of walls.

Germany In Miccoli et al. (2015), it is noteworthy that even though a first vertical crack appeared in the panels, an increase in load was still observed. Not in every sample the vertical crack was combined with the diagonal crack system.

The sample yields the stress until a first vertical crack appears by exceeding the maximum elastic horizontal strain. The crack ran not only along joints but also through blocks. At the second yielding point the sample failed by sliding of the blocks along the joints until complete collapse.

Spain Figure 18 shows the cracks pattern obtained in a diagonal compression test conducted by Rodríguez-Mariscal and Solís (2020). The typical failure corresponds to a shear stress state of the wall, appearing a diagonal crack pattern between the two loaded corners. The crack propagates mainly through the joints of the masonry, but some adobe blocks are also cracked. This indicates a very similar strength of both blocks and mortar (they are actually made of the same material in this case).

Fig. 18 Crack patterns (Rodríguez-Mariscal and Solís 2020)

3.3 *Joint Shear Behaviour*

3.3.1 Adobe Masonry Specimens: Materials and Geometry

Regarding the study joint shear behaviour, studies were made (Fontana et al. 2018) using earth blocks of size 240 mm × 115 mm × 71.5 mm with bulk density of 1809 kg/m^3 and earth mortar. Triplets for shear tests were produced. The purely mineral mortar made of natural earth and sand 0/2 mm was made of soil with 4% gravel, 75% sand and 21% clay and silt. It had a bulk density of 1880 kg/m^3, a compressive strength of 2.93 MPa and a flexural strength of 1.03 MPa. The triplets were made of three blocks mutually connected with a layer of 15 mm of earth mortar. The influence of pre-wetting the joint-contact surfaces of the blocks was assessed by submerging the blocks into water of 5 mm height for 10 s. Immediately after building, the triplets were pre-compressed with a load of 7.6 kg and were left to dry at 23 °C and 50% relative humidity for at least 28 days.

In another study (Miccoli et al. 2014), earth blocks of size 240 mm × 115 mm × 71.5 mm with bulk density of 1863 kg/m^3 and earth mortar with particle size up to 4 mm were used to produce the triplets for the shear tests. The mortar was made of soil with 55% gravel and sand, 37% silt and 14% clay. It had a bulk density of 1880 kg/m^3, a compressive strength of 3.32 MPa and a flexural strength of 1.39 MPa. The triplets were made of three blocks mutually connected with a layer of 20 mm of earth mortar and were left to dry at 23 °C and 50% relative humidity for at least 28 days.

3.3.2 Testing Procedures

In Fontana et al. (2018) and in Miccoli et al. (2014), the tests were performed through the triplet test defined by EN 1052-3, procedure A (2007). In both studies, a 50 kN universal press with a pressure cell in horizontal position to adjust the pre-compression load perpendicular to the bed joints of the specimens was used. The shear load was imposed vertically onto the middle block until it sheared away from the two outer blocks.

In Fontana et al. (2018), a total of 72 triplets were tested and three steps of pre-compression load were chosen: 0.05, 0.10 and 0.20 MPa. These stress levels lay below the allowed compression stresses for load-bearing masonry made of earth blocks, according to the German guidelines for construction with earthen materials (Lehm and Regeln-Begriffe 2009). The shear tests were performed with three triplet test specimens for each step of pre-compression.

In Miccoli et al. (2014), the joint width was adjusted to 20 mm. After preparation, the specimens were stored at 23 °C and 50% relative humidity for at least 14 days until testing. Tests were performed with four different pre-compression loads: 0.05, 0.10, 0.15 and 0.20 MPa. Chosen pre-compression loads were in the range specified by DIN 18946 (2013) to exceed not the maximum compressive strengths of earth block masonry specified by the German guidelines for construction with earthen materials (Lehm and Regeln-Begriffe 2009).

3.3.3 Cohesion, Friction Angle

In Fontana et al. (Fontana et al. 2018), for each triplet, the shear strength of each specimen was calculated as the ratio between the ultimate load and the specimen areas parallel to the mortar joint. The maximum load values tend to reflect the mechanical properties of the mortars used. In order to determine the angle of internal friction, a graph of the individual shear strength was plotted against the respective pre-compression load (normal compressive stress). Subsequently, a linear regression of the points was performed. The mean initial shear strength was obtained from the interception of the line with the vertical axis (zero normal stress). The angle of internal friction is represented by the slope of the line.

The values of mean initial shear strength and angle of internal friction for the pre-wetted blocks are 0.16 MPa and 37°, respectively. The values of mean initial shear strength and angle of internal friction for the dry blocks are 0.10 MPa and 50°, respectively (Fontana et al. 2018).

In Miccoli et al. (2014), the determination of the angle of internal friction also followed the procedures described earlier. The values of mean initial shear strength and angle of internal friction are 0.018 MPa and 49°, respectively.

3.3.4 Failure Mode

In almost all cases analysed in Fontana et al. (2018), shear failure occurred at the block-mortar interface, in few cases the earth block split in two parts. This last type of failure, due to the comparable values of compressive strength shown by earth blocks and earth mortar, occurred for both block treatments used.

In Miccoli et al. (2014), failure did not occur at the joints but, instead, in almost all cases, at the mortar-block interface. That means that even though the mortar has a compressive strength equivalent to 2/3 of the earth block's compressive strength, contact between the two materials was a very weak point concerning shear behaviour.

4 Conclusions and Final Remarks

Different authors have been conducting research on the mechanical characterization of adobe masonry. The materials and geometry of the specimens, either adobe blocks, mortar or adobe walls, have significant differences resultant from the local traditions and available materials. In most cases, clayey soil and sand are used in the fabrication of the blocks, though the percentages differ. In the bonding, earth or mud mortar is used. Reinforcements with straw were also studied.

For the testing procedures of the compressive tests, the authors applied the defined in different standards as (EN 772-1 2011; EN 1052-1 1998). In general, rigid layers, using materials as neoprene and wood, were placed in both sides of the walls for a better distribution of the stress of the uniaxial compression. The great majority of the tests were displacement controlled, though a few were force controlled.

The compressive strength of the adobe blocks is highly dependent on the size, shape and manufacturing process, which results in a wide range of results obtained by the different authors; 0.51–5.21 MPa. Similarly, the tensile strength is highly variable, being between 10 and 39% of the compressive strength. For the mortar, the compressive strength varies between 0.47 and 3.78 MPa and the flexural strength has values in the range of 0.26–1.39 MPa.

In order to analyse the shear behaviour, for the testing procedure of the diagonal compression tests, two standards were used: (ASTM-E-519-10 2010; RILEM-TC 1994), and the tests were displacement controlled. The shear strength was computed following the Standards mentioned with results ranging from 0.03 to 0.21 MPa.

The computation of the Young's Modulus depends on the type of strain measurements (tangent/secant) and points considered. Some authors determined this modulus at 1/3 of the maximum stress, others at 1/2 and others between 1/3 and 2/3 of the maximum compressive strength. The results obtained ranged from 34 to 899 MPa.

Similarly, the shear modulus determination is dependent on the type of strain measurements and points considered. The range of the results obtained is from 70 to 641 MPa.

The Poisson ratio also depends on the level of the compressive stress considered, and its values are within the interval of 0.05–0.53.

In general, due to the brittle behaviour of adobe under uniaxial load, the stress-strain curves show a short phase of post peak strain softening under compression. A distinctive yielding point is also seen in the diagonal compression test stress-strain curves.

The failure mode for the compressive behaviour registered across the several authors is similar: vertical and diagonal cracking. As for the shear behaviour, the typical failure mode is a diagonal crack pattern between the two loaded corners.

CEN 2007 (2007) and DIN 2013 (2013) were used for the joint shear behaviour analysis, with different pre-compression loads. The initial shear strength ranged from 0.018 to 0.10 MPa and the internal friction varied from 49° to 50°. Shear failure occurred, in almost all cases, at the block mortar interface.

As referred, the values of the different factors analysed are dependent on several aspects as composition of the adobe blocks, its dimensions and manufacturing techniques of the blocks and of the walls. More research is required in order to adequately characterize the different adobe construction used throughout the world. It would be extremely useful to develop a correction factor that takes into account the size, materials and construction techniques in order to be able to effectively compare adobe masonry.

References

Asprone D, Parisi F, Prota A, Fenu L, Colasanti V (Apr 2016) Adobe in Sardinia. Static and dynamic behaviour of the earthen material and of adobe constructions. Brick Block Mason 821–828, https://doi.org/10.1201/b21889-102

ASTM-E-519-07 (2007) Standard test method for diagonal tension (shear) in masonry assemblages. American Society for Testing and Materials, West Conshohocken

ASTM-E-519-10 (2010) Standard test method for diagonal tension (shear) in masonry assemblages. American Society for Testing and Materials, West Conshohocken

Augenti N, Parisi F (2010) Constitutive models for tuff masonry under uniaxial compression. J Mater Civ Eng 22(11):1102–1111

Blondet M, Vargas-Neumann J (1978) Investigación sobre vivienda rural. Lima, Peru

Brignola A, Frumento S, Lagomarsino S, Podestá S (2009) Identification of shear parameters of masonry panels through the in-situ diagonal compression test. Int J Archit Herit 3(1):52–73. https://doi.org/10.1080/15583050802138634

Caporale A, Parisi F, Asprone D, Luciano R, Prota A (2015) Comparative micromechanical assessment of adobe and clay brick masonry assemblages based on experimental data sets. Compos Struct 120:208–220. https://doi.org/10.1016/j.compstruct.2014.09.046

DIN 18945 (2013) Earth blocks—terms and definitions, requirements, test methods. German Institute for Standardisation (Deutsches Institut für Normung), Berlin

DIN 18946 (2013) Earth masonry mortar—terms and definitions, requirements, test methods. German Institute for Standardisation (Deutsches Institut für Normung), Berlin

EN 1015-11 (1999) Methods of test for mortar for masonry—part 11: determination of flexural and compressive strength of hardened mortar. European Commitee for Standardization (CEN), Brussels, Belgium

EN 1052-1 (1998) Methods of test for masonry—part 1: determination of compressive strength. European Commitee for Standardization (CEN), Brussels, Belgium

EN 1052-3 (2007) Methods of test for masonry—part 3: determination of initial shear strength. European Commitee for Standardization (CEN), Brussels, Belgium

EN 772-1 (2011) Methods of test for masonry units (part 1: determination of compressive strength). European Commitee for Standardization (CEN), Brussels, Belgium

Fontana P, Grünberg U, Miccoli L (2018) Experimental investigations on the initial shear strength of masonry with earth mortars. Int J Mason Res Innov 3(1):34–49. https://doi.org/10.1504/ijmri.2018.10009831

Illampas R, Ioannou I, Charmpis DC (2017) Experimental assessment of adobe masonry assemblages under monotonic and loading–unloading compression. Mater Struct Constr 50(1): 1–18. https://doi.org/10.1617/s11527-016-0952-z

ISO 14688-2:2004 (2004) Geotechnical investigation and testing—identification and classification of soil—part 2: Principles for a classification. International Organization for Standardization (ISO), Geneva

Lehm D, Regeln-Begriffe L (2009) Baustoffe, Bauteile, 3rd revise. Vieweg-Teubner, Wiesbaden

McNary WS, Abrams DP (1985) Mechanics of masonry in compression. J Struct Eng 111(4): 857–870

Meli R (2005) Experiencias en México sobre reducción de vulnerabilidad sísmica de construcciones de adobe. In: Proceedings of SismoAdobe2005: international seminar on architecture, construction and conservation of earthen buildings in seismic areas, Pontifical Catholic University of Peru, Lima, Peru

Miccoli L, Müller U, Fontana P (2014) Mechanical behaviour of earthen materials: a comparison between earth block masonry, rammed earth and cob. Constr Build Mater 61:327–339. https://doi.org/10.1016/j.conbuildmat.2014.03.009

Miccoli L, Garofano A, Fontana P, Müller U (2015) Experimental testing and finite element modelling of earth block masonry. Eng Struct 104:80–94. https://doi.org/10.1016/j.engstruct.2015.09.020

Müller P, Miccoli L, Fontana P, Ziegert C (2017) Development of partial safety factors for earth block masonry. Mater Struct Constr 50(1):1–14. https://doi.org/10.1617/s11527-016-0902-9

NMX-C-085-ONNCCE-2002 (2002) Industria de la construcción-Cementos hidráulicos-Método estándar para el mezclado de pastas y morteros de cementantes hidráulicos. México DF: Organismo Nacional de Normalización y Certificación para la Construcción y Edificación (in Spanish)

Norma E.080 (2017) Diseño y construcción con tierra reforzada, Ministerio de Vivienda, Construcción y Saneamiento (MVCS), Lima (in Spanish)

NZS 4210 (2001) Masonry construction: materials and workmanship. Standards New Zealand (SNZ), Wellington

NZS 4298 (1998) Materials and workmanship for earth buildings. Standards New Zealand (SNZ), Wellington

Parisi F, Iovinella I, Balsamo A, Augenti N, Prota A (2013) In-plane behaviour of tuff masonry strengthened with inorganic matrix-grid composites. Compos Part B Eng 45(1):1657–1666. https://doi.org/10.1016/j.compositesb.2012.09.068

Parisi F, Asprone D, Fenu L, Prota A (2015) Experimental characterization of Italian composite adobe bricks reinforced with straw fibers. Compos Struct 122:300–307. https://doi.org/10.1016/j.compstruct.2014.11.060

RILEM-TC (1994) Diagonal tensile strength tests of small wall specimens. RILEM Recommendations for the Testing and Use of Constructions Materials

RILEM-TC-76-LUM (1994) Diagonal tensile strength tests of small wall specimens. International Union of Laboratories and Experts in Construction Materials, Systems and Structures, Bagneux

Rodríguez-Mariscal JD, Solí M (2020) Hacia una metodología para la caracterización experimental del comportamiento a compresión de la mampostería de adobe. Inf la Construcción 72(557):332 https://doi.org/10.3989/ic.67456 (in Spanish)

San Bartolomé A, Pehovaz R (2005) Comportamiento a carga lateral cíclica de muros de adobe confinados. In: SismoAdobe2005 Conference

Silveira D, Varum H, Costa A, Carvalho J (2015) Mechanical properties and behavior of traditional adobe wall panels of the Aveiro district. J Mater Civ Eng 27(9):04014253. https://doi.org/10.1061/(asce)mt.1943-5533.0001194

Tarque N, Camata G, Spacone E, Varum H, Blondet M (2014a) Numerical simulation of an adobe wall under in-plane loading. Earthquakes Struct 6(6):627–646

Tarque N, Camata G, Spacone E, Varum H, Blondety M (2014b) Non-linear dynamic analysis of a full-scale unreinforced adobe model. Earthq Spectra J 30(4):759–794

Torrealva D, Acero J (2005) Reinforcing adobe buildings with exterior compatible mesh. The final solution against the seismic vulnerability? In: SismoAdobe2005: international seminar on architecture, construction and conservation of earthen buildings in seismic areas

Vargas-Neumann J, Ottazzi G (1981) Investigaciones en adobe. Lima, Peru

Vasconcelos G, Lourenço PB (2009) Experimental characterization of stone masonry in shear and compression. Constr Build Mater 23(11):3337–3345. https://doi.org/10.1016/j.conbuildmat.2009.06.045

Walker P (2002) The Australian earth building handbook, HB 195-2002. Standards Australia, Sydney

Wu F, Li G, Li HN, Jia JQ (2013) Strength and stress-strain characteristics of traditional adobe block and masonry. Mater Struct 46(9):1449–1457. https://doi.org/10.1617/s11527-012-9987-y

Yamin LE, Phillips C, Reyes JC, Ruiz D (2007) Estudios de vulnerabilidad sísmica, rehabilitación y refuerzo de casas en adobe y tapia pisada. Apunt Rev Estud sobre Patrim Cult 20(2):286–303 [Online]. Available: c:%5CUsers%5CUsuario%5CDesktop%5CPhD%5CPapers%5CRammed earth Papers%5C; Lacouture L.E.Y. et al—Estudios de vulnerabilidad sismica, rehabilitacion y refuerzo de casas en adobe y tapia pisada.pdf

Quasi-static In-Plane Testing of Adobe Masonry Walls and Structures

Nicola Tarque, Fulvio Parisi, Domenico Asprone, Andrea Prota, Dora Silveira, Marcial Blondet, and Humberto Varum

Abstract The lateral seismic capacity and the in-plane failure pattern of adobe masonry may be understood by performing quasi-static experimental tests on full-scale adobe walls. Information regarding the maximum lateral capacity and the displacement ductility are normally obtained during these tests. Also, information about limit states could be obtained. Along the years, universities in Europe, Asia and Latin America have performed this type of tests and the results of some of these works are summarized in this chapter.

Keywords Cyclic behaviour · Adobe wall · In-plane capacity · Failure pattern

N. Tarque (✉) · M. Blondet
GERDIS Research Group, Civil Engineering Division, Department of Engineering, Pontificia Universidad Católica del Perú, Av. Universitaria 1801, San Miguel, Lima, Peru
e-mail: sntarque@pucp.edu.pe

M. Blondet
e-mail: mblondet@pucp.edu.pe

F. Parisi · D. Asprone · A. Prota
Department of Structures for Engineering and Architecture, University of Naples Federico II, Via Claudio 21, 80125 Naples, Italy
e-mail: fulvio.parisi@unina.it

D. Asprone
e-mail: d.asprone@unina.it

A. Prota
e-mail: aprota@unina.it

D. Silveira · H. Varum
CONSTRUCT-LESE, Civil Engineering Department, Faculty of Engineering, University of Porto, R. Dr. Roberto Frias, 4200-465 Porto, Portugal
e-mail: dsilveira@fe.up.pt

H. Varum
e-mail: hvarum@fe.up.pt

H. Varum et al. (eds.), *Structural Characterization and Seismic Retrofitting of Adobe Constructions*, Building Pathology and Rehabilitation 20,
https://doi.org/10.1007/978-3-030-74737-4_5

1 Introduction

The objective of the tests addressed in this chapter is the evaluation of the seismic in-plane behaviour of adobe walls through the evaluation of the force versus horizontal curve and the evolution of diagonal cracks at different levels of horizontal top displacement. Normally, the incremental loading is applied at low rates and the top wall part. The slow velocity is to avoid the contribution of any inertia force (i.e. acceleration). The internal forces are given by the restitutive and the damping forces and the external force is given just by the applied lateral load. The wall could be also preliminary subjected to constant vertical loading.

In Fig. 1 a scheme of the in-plane test is showed. Here the lateral load V causes the top displacement Δ, producing shear and bending deflections in the entire wall. However, bending deflections are almost null since those are around 1% of the shear deflections. Then, the lateral stiffness of the wall is given by $K = V/\Delta$, computed in the elastic range. Special attention is also needed to understand the boundary conditions of the test. Since the type of failure depends a lot on boundary conditions, special attention is needed during the test set up. The types of failure that are presented in adobe walls are sliding, shear and bending, but normally a mix of them occur in seismic loading.

2 Test Performed at the Pontifical Catholic University of Peru

The in-plane cyclic behaviour of three I-shape adobe walls without reinforcement was studied by Blondet et al. (2005, 2008), and those are explained as follows.

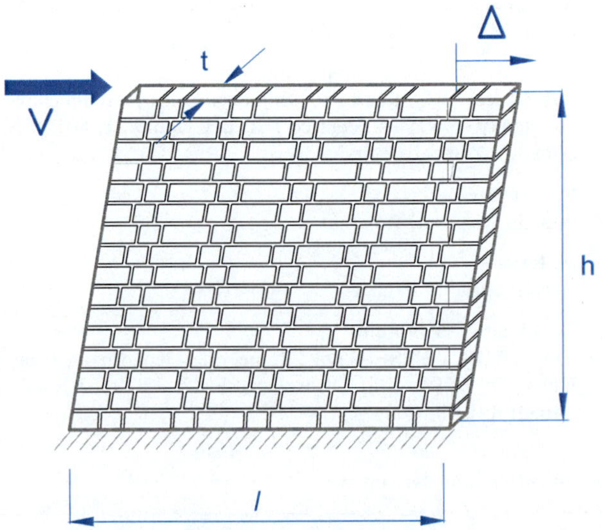

Fig. 1 Lateral deformation of an adobe wall due to shear forces

2.1 Wall Characteristics

Here the walls are identified as Wall I-1 (Fig. 2a), Wall I-2 and Wall I-3 (Fig. 2b). The geometrical characteristics were the same for all of them; the only difference was in Wall I-1, which had a 0.40 × 0.60 m central window opening. The main longitudinal wall was 3.06 m long, 1.93 m high and 0.30 m thick. All walls had two 2.48 m transverse walls (Fig. 3). The geometrical configuration of Wall I-2 and I-3 is the same as presented in Fig. 3 but without window opening. The adobe bricks used for the wall construction had dimensions of 0.13 × 0.10 × 0.30 m and 0.13 × 0.10 × 0.22 m. The brick composition for all walls was soil, coarse sand and straw in proportion 5/1/1, and for the mud mortar, 3/1/1. The bricks were laid alternating headers and stretchers in the courses.

Each specimen was built over a reinforced concrete foundation beam. At the top, a reinforced concrete crown beam was built to provide the gravity loading corresponding to the roof of a typical Peruvian dwelling consisting of wooden beams, cane, straw, mud and corrugated zinc sheets. The total weight of each wall was approximately 135 kN, which considers the weight of the concrete beams. The top concrete beam had a 16 kN weight, the foundation beam had a 31.40 kN weight, and the adobe wall had 87.6 kN weight for the I-1 wall and 89.0 kN for the other two walls. The lintel in Wall I-1 was made of wood.

The load was applied horizontally at the top concrete beam through a servo-hydraulic actuator with a maximum capacity of 500 kN, placed on a rigid steel frame. To avoid a concentrated load, steel and wooden plates were placed at the contact of the actuator with the crown beam (Fig. 4a). Besides, two steel rods were placed at the wall top to improve the horizontal load transmission all over the wall, to simulate a distributed horizontal load. A total of 17 Linear Variable Displacement Transducer (LVDT) was placed in the wall. One side of the wall was painted white to help the cracks visualization.

(a) With window opening (1 specimen)

(b) Without window opening (2 specimens)

Fig. 2 Adobe walls subjected to the cyclic test (Blondet et al. 2005)

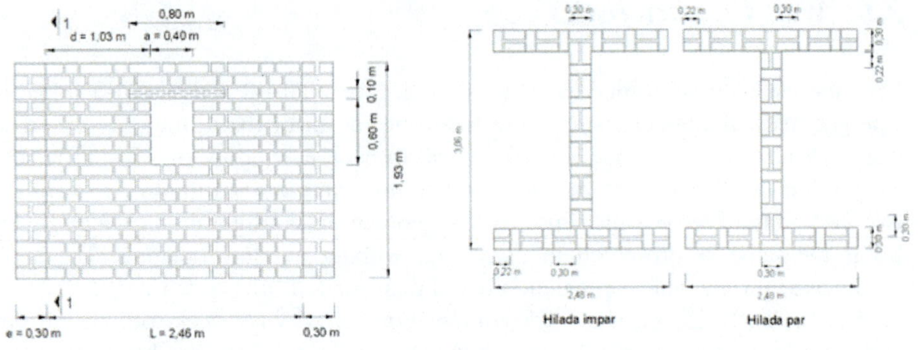

(a) Front view (b) Plan view

Fig. 3 Scheme of the adobe walls for the cyclic test (Blondet et al. 2005)

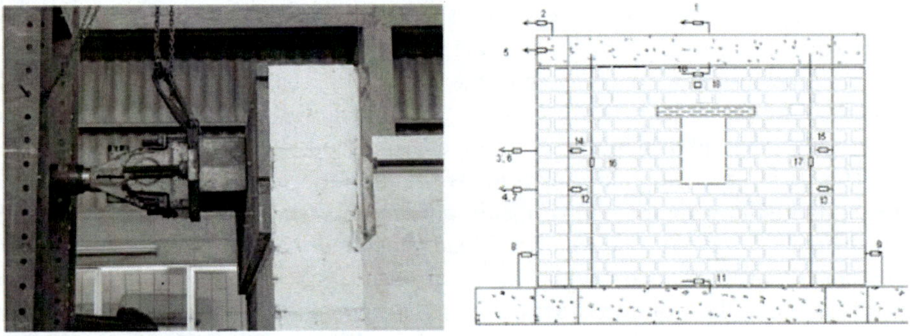

(a) Contact of the actuator with the wall (b) Set-up of instrumentation on the wall

Fig. 4 Load application and instrumentation on the adobe walls (Blondet et al. 2005)

2.2 Experimental Tests

The tests were displacement controlled. The load application velocity was incremented in each phase as reported in Table 1. The cyclic load was applied in 7 phases (push and pull), with 2 cycles for each phase. The maximum displacements, incremented in each phase, were 0.1, 0.5, 1, 2, 5, 10 and 20 mm.

The crack evolution was quite similar for all the walls as described in Table 1. The first phase was useful for calibration of the instrumentation. The diagonal fissures (x-shape) started to appear during the third and fourth phases, with loss of strength. In the fifth phase, large horizontal fissures appeared at the transversal walls and vertical fissures at the intersection of longitudinal and transversal walls, with an increment of diagonal cracking in the main wall. During the sixth phase (corresponding to 10 mm as top displacement) a notable loss of load strength in the walls was observed with an increment of crack thickness and tensile cracking in the adobe

Table 1 Description of damage on the walls subjected to the cyclic tests

Phase	Wall ID	Δ (mm)	Velocity (mm/min)	Maximum load (kN)	Crack thickness (mm)
1	I-1	0.1	0.1	–	–
	I-2			–	–
	I-3			–	–
2	I-1	0.5	0.5	26	–
	I-2			36	–
	I-3			42	–
3	I-1	1	1.0	32	0.60
	I-2			49	0.10
	I-3			52	0.25
4	I-1	2	2.0	38	0.60
	I-2			48	0.50
	I-3			53	0.50
5	I-1	5	5.0	38	4.0
	I-2			40	2.5
	I-3			41	2.5
6	I-1	10	10.0	34	10.0
	I-2			43	6.0
	I-3			39	6.0
7	I-1	20	20.0	32	50.0
	I-2			44	15.0
	I-3			39	15.0

bricks. The diagonal cracking in both directions continued growing in thickness. Horizontal cracks appeared at the base of the transversal walls, allowing a sliding behaviour of the walls. At this stage, some rigid blocks were identified. The last phase showed a complete loss of load strength with the formation of cracks that cut the adobe bricks in the principal and transversal walls. The last phase involved sliding of the walls with greater crack width.

2.3 *Result's Discussion*

All the walls had a similar in-plane behaviour: X-shaped diagonal cracks, horizontal cracks in the transversal walls, and loss of strength after 2 mm top displacement. The two walls without windows had larger initial stiffness than the wall with opening, as seen in the hysteretic curves in Fig. 5. In terms of lateral force, Wall I-2 and Wall I-3 resisted around 25% more than Wall I-1. The last two walls were used for another research program to investigate the effect of crack grouting. The injected cracks in the Wall I-2 and I-3 are identified in the left photos in Fig. 5.

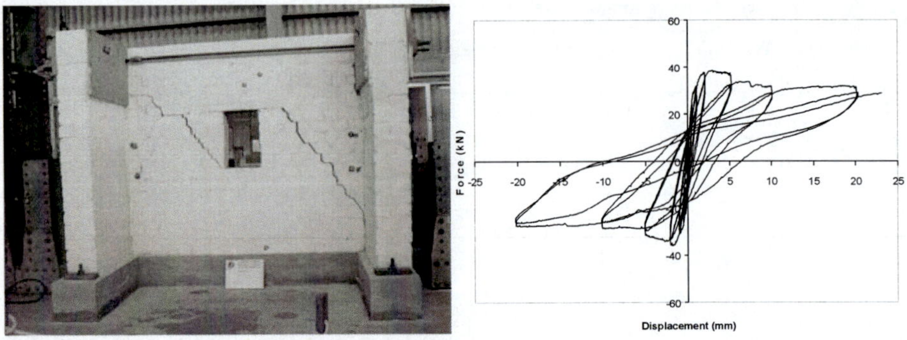

(a) Crack pattern and load-displacement response, wall I-1

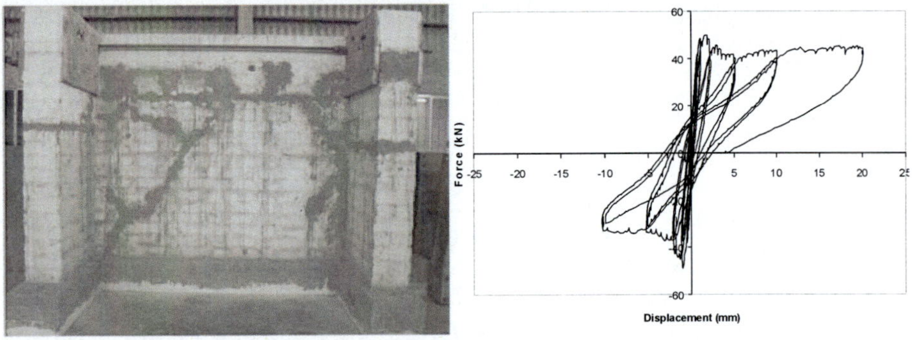

(b) Crack pattern and load-displacement response, wall I-2

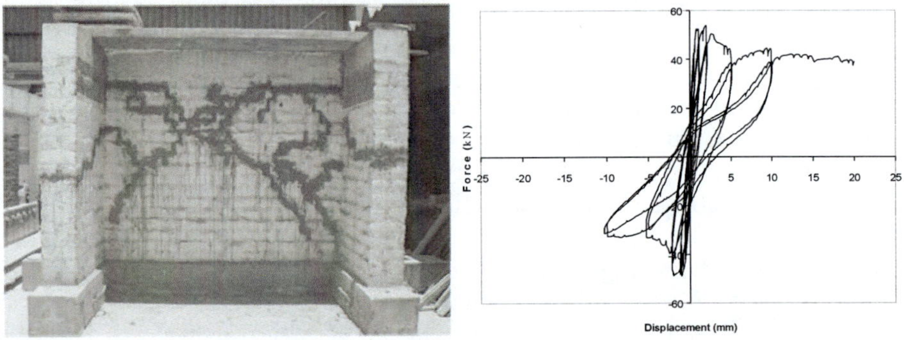

(c) Crack pattern and load-displacement response, wall I-3

Fig. 5 Hysteretic curves and crack pattern of the adobe walls subjected to the cyclic tests. All the figures on the *left* refer to the end of the cyclic tests (Blondet et al. 2005, 2008)

It seems that at the beginning of loading there is an adjustment of the hydraulic actuator and instrumentation, giving an apparent initial stiffness greater than the real one. To avoid this problem, the initial stiffness should be computed from the hysteretic curves when the maximum displacement at the top is around 1 mm. Beyond this displacement, some fissures started to appear in all the walls.

The hysteretic curves yield useful results up to a 10 mm top displacement. Beyond this value sliding behaviour of the wall due to the relative movement between rigid blocks along the cracks was observed. Figure 5 shows the hysteretic curves for each wall and a picture representing the crack pattern at the end of the cyclic tests. For Wall I-1, the x-shape cracks started at the window corners and grew diagonally to the top and base of the wall. The wider cracks were formed when the wall was pushed. Unlike the previous wall, Wall I-2 and Wall I-3 were repaired and tested again; however, the responses of the repaired walls are not reported here. Figure 5b, c show the repaired crack patterns.

Figure 6 shows the envelopes of the experimental hysteretic curves, in both positive and negative directions. Since the wall was first pushed, the positive branch of each hysteretic curve is stronger than the negative branch, which represents the behaviour of the wall when it is pulled. To compute the initial part of the envelope curve, the data obtained before 1 mm top displacement was neglected since it can contain noise due to the equipment calibration.

2.4 *Evaluation of the Modulus of Elasticity E*

Some useful equations for the evaluation of the wall stiffness, considering deformations due to shear and bending, are reported as follows:

- Stiffness when the wall is double fixed:

$$K = \frac{1}{\frac{h^3}{12E \cdot I} + \frac{h}{G \cdot A_v}} \tag{1}$$

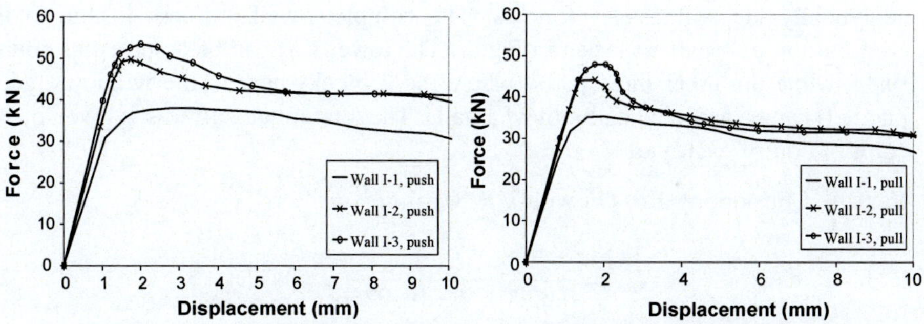

(a) Positive branch of the hysteretic curves (b) Negative branch of the hysteretic curves

Fig. 6 Envelope of the hysteretic curves (positive and negative branch) from the cyclic tests

- Stiffness for cantilever walls (one end is fixed and the other is free):

$$K = \frac{1}{\dfrac{h^3}{3E \cdot I} + \dfrac{h}{G \cdot A_v}} \tag{2}$$

- Total stiffness obtained from the contribution of walls connected in series:

$$\frac{1}{K_T} = \sum_{i=1}^{n} \frac{1}{K_i} \tag{3}$$

- Total stiffness obtained from the contribution of walls connected in parallel:

$$K_T = \sum_{i=1}^{n} K_i \tag{4}$$

In the previous equations, h is the wall height, E is the elasticity modulus, I is the area moment of inertia, G is the shear modulus taken as $0.4E$ and A_v is the effective shear area. This last value has a shear deformation factor of 1.2.

2.4.1 Wall I-1

The experimental initial stiffness K can be computed from the pushover curves showed before. For wall I-1 the initial stiffness is estimated as:

$$K_1 = \frac{F}{\Delta} \approx 28\,\text{kN/mm}$$

The influence of the central window should be taken into account for computing analytically the stiffness as follows. The complete wall stiffness is due to the contribution of 4 parts, as seen in Fig. 7. The lower part can be assumed as double fixed, while the other three parts (where the 2 blocks next to the windows are in parallel) can be assumed as cantilever walls. The total initial stiffness is given by the contribution of each part.

Part A. Considering Eq. (1) with $h = 876\,\text{mm}$

$$K_A = \frac{E}{0.0000174 + 0.00286} = 347E$$

It can be seen that the deformation due to bending is almost negligible. Therefore, considering a cantilever beam or a double fixed element will not influence the final evaluation of the initial stiffness. For Part A, considered as a cantilever beam, K_A is $341E$.

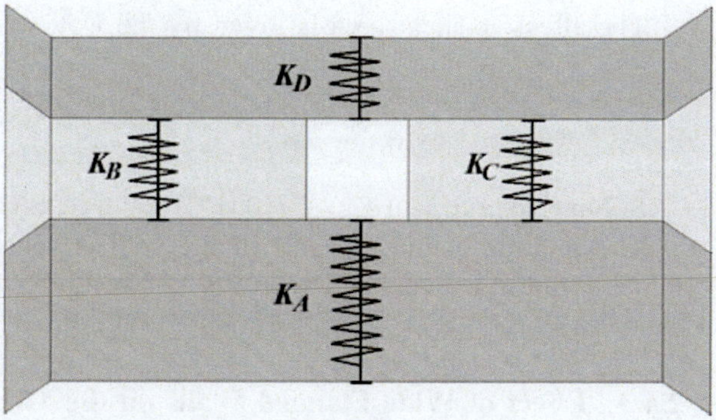

Fig. 7 Scheme for evaluating the lateral stiffness of the adobe wall I-1

Part B and Part C. Considering Eq. (2) with h = 600 mm

$$K_B = K_C = \frac{E}{0.00005564 + 0.004511} = 219.1E$$

Again, it is seen that the influence of the bending deformation is small compared with the shear deformation. For Part B and C, considered as a double fixed beam, K_B is 221E.

Part D. Considering Eq. (2) with $h =$ 454 mm

$$K_D = \frac{E}{9.87 \times 10^{-6} + 0.00148} = 669.7E$$

The total initial stiffness is computed combining Eqs. (3) and (4) as follows:

$$\frac{1}{K_T} = \frac{1}{K_A} + \frac{1}{K_B + K_C} + \frac{1}{K_D}$$

Evaluating the previous equation, the total stiffness is $K_T \approx 150E$. Then, $E \approx 187$ MPa.

2.4.2 Wall I-2 and I-3

In this case, there is not a window opening. The wall can be assumed as fixed at both ends with a distributed horizontal load at the top. The experimental initial stiffness computed from the pushover curves is around K_T = 34 kN/mm

The area moment of inertia is:

$$I = \frac{2480 \times 3060^3}{12} - 2\frac{1090 \times 2460^3}{12} = 3.217 \times 10^{12}\ \text{mm}^4$$

The effective shear area is given by the wall web:

$$A_v = \frac{5}{6}A_w = \frac{5}{6}(2460 + 600)\times 300 = 765{,}000\,\text{mm}^2$$

Solving for Eq. (1): $K = E/(0.0001862 + 0.0063) = 154E$. Then, $E \gg 220\,\text{MPa}$.

The rotation at the top part of the wall is accepted (cantilever beam); so, the initial stiffness is evaluated with Eq. (2), resulting in $E \approx 240\,\text{MPa}$.

2.4.3 Effect of Wide Flanged Walls on the Modulus of Elasticity E

Paulay and Priestley (1992) suggest computing an effective width on wide- flanged walls to evaluate the seismic capacity of a wall subjected to shear forces, as seen in Fig. 8. This is based on the assumption that vertical forces due to shear stresses introduced by the web of the wall into the tension flange spread out at a slope of 1:2.

According to Fig. 8, the effective width of the tension flange is expressed as:

$$b_{eff} = h_w + b_w \leq b \tag{5}$$

where h_w is the wall height, b_w is the in-plane wall width and b is the length of the flange wall. The effective width in compression is given by:

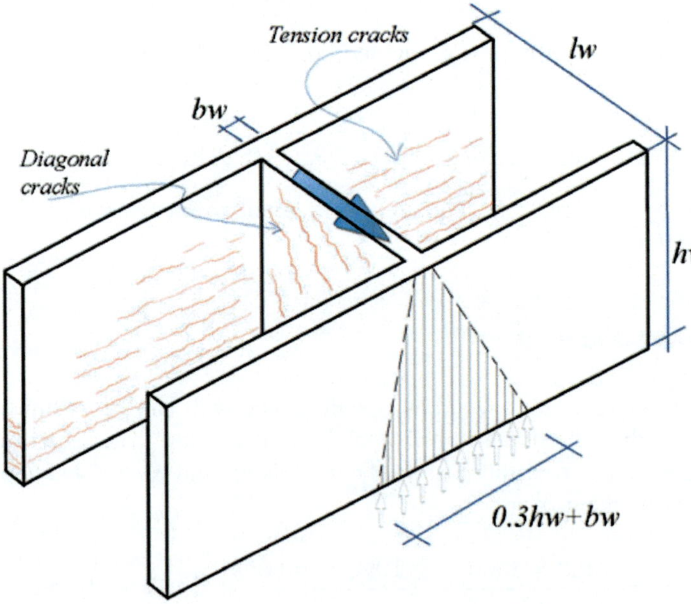

Fig. 8 Estimation of effective flange widths in structural walls, modified from Paulay and Priestley (1992)

$$b_{eff} = 0.3h_w + b_w \leq b \tag{6}$$

With the two expressions written above, the effective widths for the flanges are 2,230 and 879 mm for the tension and compression zone, respectively.

The coordinate of the gravity centre along the application of the force is:

$$\bar{y} = \frac{2230 \times 300 \times 150 + 2460 \times 300 \times 1530 + 879 \times 300 \times 2910}{2230 \times 300 + 2460 \times 300 + 879 \times 300}$$
$$= 1,189.22\,\text{mm}$$

The area moment of inertia is $I_a = 1.97 \times 10^{12}\ \text{mm}^4$.

The effective shear area is given by the wall web:

$$A_v = \frac{5}{6}A_w = \frac{5}{6}(2460 + 600) \cdot 300 = 765,000\,\text{mm}^2$$

Solving for Eq. (2): $K = E/(0.0003 + 0.006307) = 151E$. Then, $E \approx 224\,\text{MPa}$.

It is seen that the analytical computed elasticity values for the elasticity modulus are around 220 MPa. Some other researches (Tarque et al. 2014) have shown high variability in the elasticity modulus, varying from 100 to 250 MPa, but they suggest using values around 170 MPa. The elasticity modulus is more sensitive to shear deformation than flexural deformation. As previously discussed, it seems that there is no major difference in assuming a double fixed wall or a cantilever wall. From the analysis presented here, it can be concluded that a E = 200 to 220 MPa (Tarque et al. 2014) can be used for the numerical analysis.

2.5 *Capacity Curve for Adobe Walls*

To obtain the capacity curve of an adobe wall, the hysteretic curve and the damage pattern of Wall I-1 have been analysed (Fig. 5a). The envelope (maximum force values) and the drift limits are shown in Fig. 9. These drift limits are obtained by analysing the cracking progress during the test. Until 0.05% (drift) the structure can be considered as elastic (LS1), which means fully operational. After that, the structure can have some cracks but is still functional (LS2) until 0.1% of drift. Then the life-safety performance (LS3) is reached at 0.26% of drift and finally, the structure is considered near to collapse or collapsed at 0.52% of drift.

According to the previous analysis, four LS have been adopted for 1-story adobe masonry buildings and these are described as follows:

- **LS1**: Only very minor damage has occurred. The building retains its original stiffness and strength. The behaviour of the building is essentially elastic and stable. The risk of life-threatening injury during the earthquake is negligible.

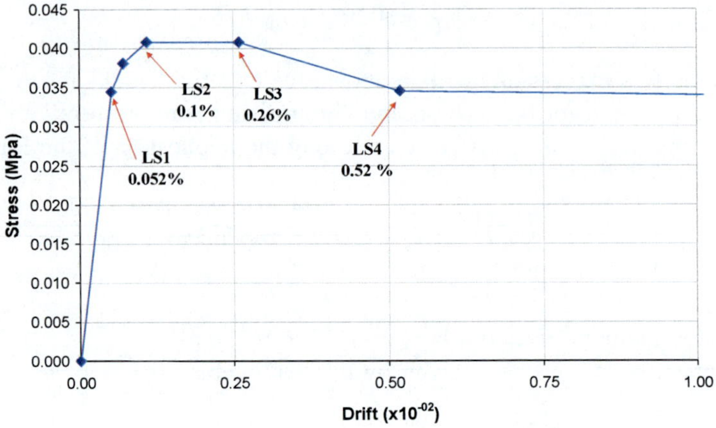

Fig. 9 Capacity curve for an adobe wall subjected to an in-plane force

- **LS2**: Minor structural damage can be seen as slightly diagonal cracks. The structure retains nearly all its original stiffness and strength. Repairs may be instituted at the convenience of the building users. The risk of life-threatening injury during the earthquake is very low.
- **LS3**: Significant structural damage is reported. Beginning of horizontal cracks. Some adobe bricks have been cut. The building has lost a significant amount of its original stiffness but retains some lateral strength and margin against collapse. The building cannot be used after the earthquake without significant repair.
- **LS4**: Near collapse or collapse of the building is expected. Repairing the building is neither possible nor economically reasonable. The structure will have to be demolished after the earthquake. Beyond this LS, global collapse with danger for human life has to be expected.

Similar to Calvi (1999), five situations can result from the comparison of expected demand and the capacities discussed above.

Nearly undamaged building	Demand < LS1
Slightly diagonal cracks, usable building	LS1 < Demand < LS2
Building extensively damaged but still reparable. Beginning of horizontal cracks	LS2 < Demand < LS3
Buildings not collapsed but with severe damaged. Continuation of horizontal cracks. Cutting of adobe blocks	LS3 < Demand < LS4
Collapsed buildings	Demand > LS4

2.6 Evaluation of the Equivalent Viscous Damping in Adobe Walls

The equivalent viscous damping ratio for adobe walls is calculated considering the energy absorbed in the hysteretic loop (steady-state cyclic response) due to a given displacement level. In this case, the equivalent viscous damping for wall *I*-1 is evaluated for each limit-state with the following equation:

$$\xi_{hyst} = \frac{A_h}{4 \cdot \pi \cdot A_e} \tag{7}$$

where A_h is the area within a complete cycle of stabilized force-displacement response, and A_e is the elastic area.

Typically, the dissipated energy in each cycle evolves with the increase of damage and with the increase of displacement demand. In the work developed by Magenes and Calvi (1997), equivalent damping ratios for masonry walls were evaluated from cyclic and pseudo-dynamic tests and considering a flexural response, diagonal shear cracking response and shear sliding response. For the shear cracking response, it is adverted that greater values of equivalent damping ratios can be obtained if a cyclic test is considered instead of the pseudo-dynamic one, with the ratio of the two values being around 1.34. Besides, if it is considered the first cycle, a very high amount of hysteretic energy is dissipated, while the remaining cycles tend to show lower dissipation. Figure 10 shows the damping ratios obtained from masonry wall tests (Calvi 1999). As it can be seen, the equivalent damping ratios suggested for LS1 and LS2 (0.1% drift), and LS3 (0.3% drift) are less than those obtained from the masonry hysteretic curves. While for the LS4 of masonry, the damping ratio is related to the first cycle of the test. Looking into these aspects and knowing the limitations to infer in damping values from just one cyclic test, the average equivalent damping ratios for adobe walls has been adjusted to 10%, 10%, 12% and 16% for LS1, LS2, LS3 and LS4 respectively (Table 3).

Fig. 10 Equivalent hysteretic damping from the cyclic test for masonry and adobe walls, adapted from Calvi (1999)

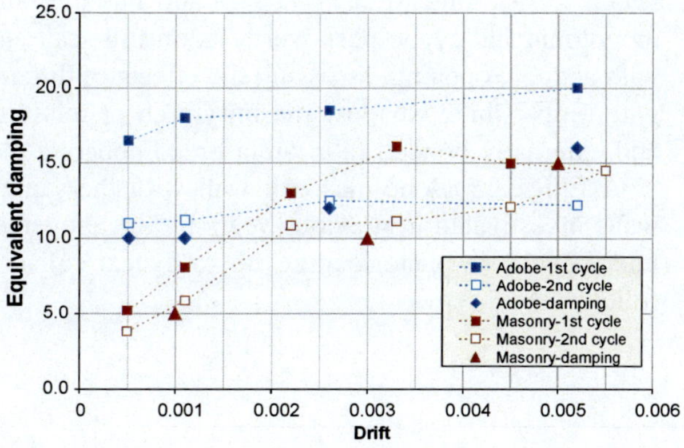

Table 2 Damping ratios for adobe walls

Limit state	Drift (%)	ξ (%)		
		1st cycle	2nd cycle	Average
LS1	0.052	16.5	11.0	13.8
LS2	0.110	18.0	11.2	14.6
LS3	0.260	18.5	12.5	15.5
LS4	0.520	20.0	12.2	16.1

Table 3 Limit states for adobe buildings

Limit state	Description	Drift (%)	ζ (%)	Ductility
LS-1	Operational	0.052	10	1
LS-2	Functional	0.10	10	2
LS-3	Life-safety	0.26	12	5
LS-4	Near or collapsed	0.52	16	10

Then, for the wall *I*-1 the equivalent damping was computed evaluating the energy dissipated by the first and second cycle of each hysteretic curve. The resulting values related to each of the drift limits clearly shows that damping evaluated in the first cycle are higher than the ones of the second cycle (Table 2).

Following the results and conclusions of Magenes and Calvi (Calvi 1999), the damping ratios showed in Table 3 are recommended for unreinforced adobe walls.

3 Test Performed at University of Naples Federico II

The seismic capacity of masonry walls under in-plane lateral actions strongly depends on the presence, distribution and size of openings. In such a case, spandrels above openings provide coupling between piers, influencing the distribution of lateral loads. Augenti et al. (2011) and Parisi et al. (2014) carried out a series of monotonic and cyclic pushover tests on full-scale, tuff stone masonry (TSM) walls with a single opening, assessing the effects of different spandrel configurations (i.e. with timber lintel, shallow masonry arch or reinforced concrete (RC) bond beam) and externally bonded fabric-reinforced concrete matrix (FRCM) systems.

A full-scale adobe masonry wall with the same overall geometry of the TSM walls investigated in previous years by the abovementioned researchers was tested under cyclic in-plane loading up to a near-collapse stage and it is described as follows.

3.1 Wall Characteristics

Figure 11 shows the specimen geometry and test set-up. The total height and length of the wall were equal to 3.60 and 5.10 m, respectively, whereas the thickness was equal to 0.40 m. The opening was located in the centre of the wall and was 1.70 m-long and 2.00 m-high. The wall was built on a rigid RC basement that was anchored to the laboratory strong floor using steel bars, plates and bolts. A rigid steel I-beam was placed on top of each pier to evenly distribute the vertical load, which was applied by a single hydraulic jack. To that aim, a mortar layer was realised on top of both piers. The reaction system of each hydraulic jack consisted of three main components: a rigid steel I-beam running in the direction perpendicular to the wall, and a couple of steel bars. The latter anchored the reaction beam to the RC basement of the wall and were hinged at their ends to accommodate any lateral displacement and rotation of each pier, avoiding a parasite contribution of the test set-up to the overall response of the specimen. For similar purposes, a spherical hinge was placed between each hydraulic jack and the overlying reaction beam. A single servo-hydraulic actuator anchored to a vertical RC reaction wall was installed to apply horizontal loading. A cylindrical hinge and a rigid steel plate at the end of the actuator ensured a conservative loading system, i.e. horizontal loading independent of the actual deformed shape. The end plate of the actuator was perforated to allow the installation of three horizontal steel bars running on each side of the wall, along the whole length of the spandrel. On the opposite edge of the wall, those bars were fixed to another perforated steel plate to allow the application of cyclic load reversals.

Adobe bricks were reinforced with a random distribution of straw fibres and were 100 mm^3 × 200 mm^3 × 400 mm^3 in size. Details on their physical and mechanical properties can be found in Chap. 3, according to the outcomes of an experimental program by Parisi et al. (2015). Mortar joints of the adobe wall were 10 mm-thick and had the same composition of bricks without straw fibre

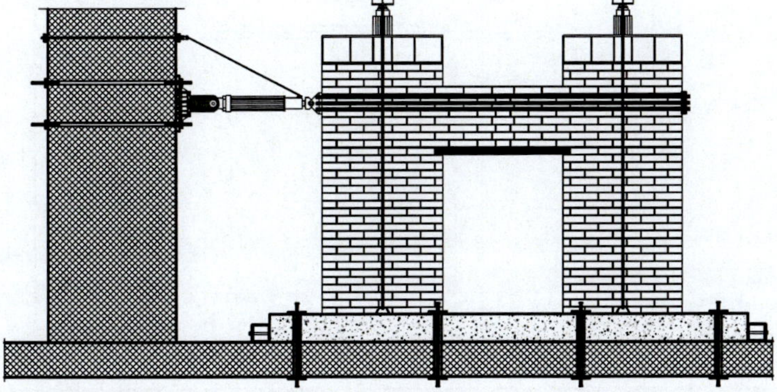

Fig. 11 Experimental set-up of an adobe masonry wall with an opening

reinforcement. The adobe masonry of the wall specimen was built according to a running bond scheme, with transverse bricks only at pier edges to simulate a poor construction quality.

Figure 12 shows the instrumentation applied over the front and back sides of the wall specimen. Linear variable differential transformers (LVDTs) were installed at the location of critical end sections of the piers and central spandrel element, to measure axial displacements due to bending.

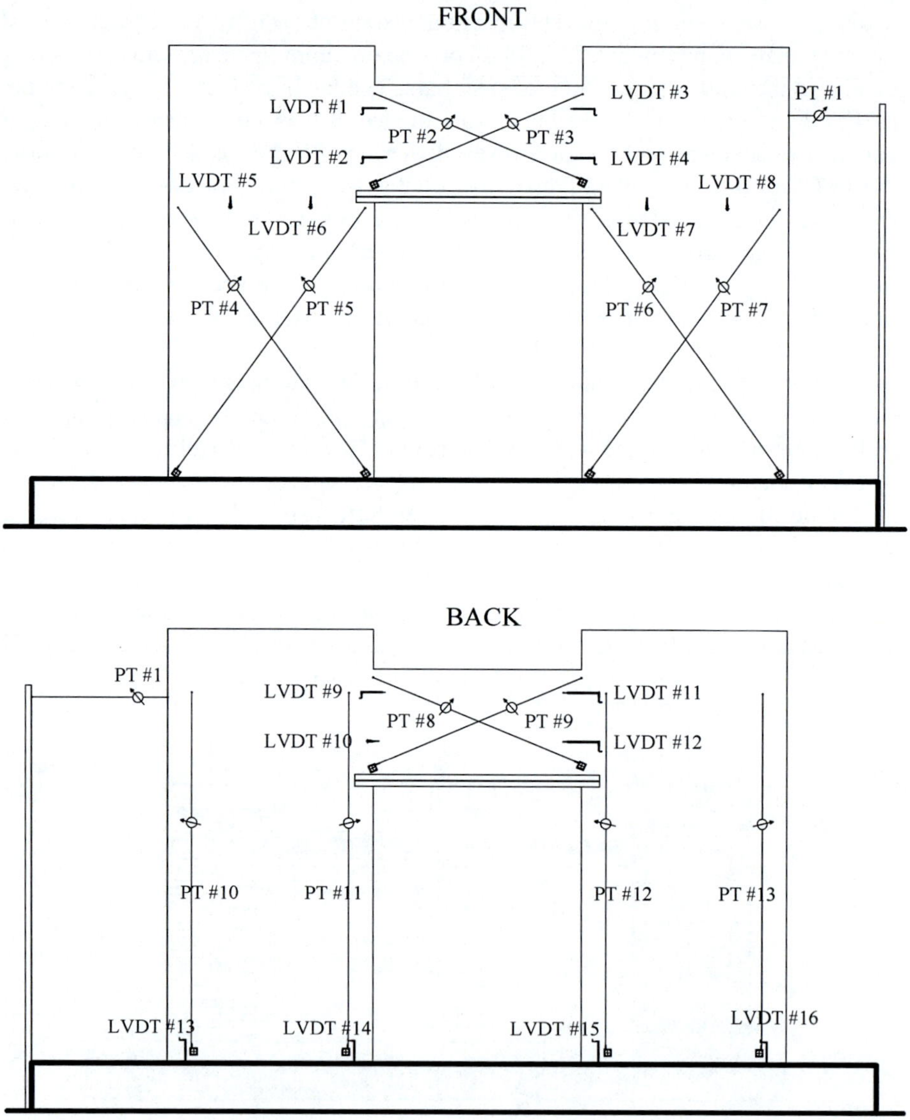

Fig. 12 Instrumentation of adobe masonry wall with an opening

On the backside of the specimen, vertical LVDTs at the base of the piers allowed the authors to measure the width of rocking-induced cracks, which produce large horizontal displacements. Those displacement transducers were complemented by two couples of vertical wire potentiometers (PTs) that measured the axial and bending deformations of each pier. Shear deformations were monitored via diagonal PTs attached to both piers and spandrel panel above the opening.

3.2 Experimental Test

The experimental test was performed in a quasi-static fashion with displacement control to evaluate all features of the in-plane seismic behaviour of the wall specimen (e.g. strength and stiffness degradation). The loading protocol of the test consisted of two stages. The former was the application of vertical loads with force control on top of piers to simulate gravity loads transferred by an ideal floor. After that two load cycles were completed to obtain effective contrast between jacks and piers, the axial load on piers was monotonically increased according to a force rate equal to 1 kN/s until a maximum axial load of 88 kN was reached (corresponding to 10% of the mean ultimate load of pier cross-sections). In the second stage of the test, the axial load on piers was kept constant and in-plane horizontal loading was imposed on top of the spandrel. Horizontal loading was applied following a target displacement time history, which was composed of four cyclic displacement blocks, each of them consisting of three cycles at each amplitude peak. The displacement rate was set to 0.70 mm/s, whereas the displacement increment between consecutive blocks of displacement cycles was equal to 5.6 mm. Thus, a total number of twelve displacement cycles were imposed on the specimen, reaching a maximum displacement of 33.6 mm. Figure 13 illustrates the wall specimen before and after the test. The sampling rate of all instruments was 5 Hz, allowing good monitoring of deformations during the entire test.

(a) Before the test (b) After the test

Fig. 13 Adobe masonry wall with an opening

3.3 Result's Discussion

Figure 14 shows the crack patterns observed on the specimen at the end of the experimental test. Whilst moderate damage to the spandrel panel was observed, the piers suffered large cracks. The spandrel failed in shear, evidencing horizontal cracks along different bed joints. By contrast, the piers experienced a bending failure with horizontal cracks concentrated in the first six-to-seven masonry layers from the base. In Fig. 16, the horizontal force recorded by a load cell of the actuator is plotted against the horizontal displacement measured by the wire potentiometer PT#1, which was positioned at the same height and connected to a wall edge (see Fig. 12). The maximum horizontal force was equal to 86.4 kN in the positive orientation associated with a pushing action of the actuator, whereas a peak force level equal to 81.4 kN was recorded in the opposite orientation. The test was stopped when a horizontal force of 61.4 kN was attained on the last positive displacement cycle, evidencing approximately a 25% reduction in lateral resistance. In that condition, the large cracks observed at the base of the piers led the specimen to be strongly unstable, highlighting out-of-plane deformations. Apart from the effect of wall self-weight, the lateral resisting force (it is average between those related to both orientations of loading) was equal to approximately 31% of vertical load. Such a ratio indicates an in-plane acceleration capacity of 0.31 g.

It is worth noting that the TSM wall with the same spandrel configuration tested by Augenti et al. (2011) showed an acceleration capacity equal to 0.38 g, which is 1.2 times the lateral load-bearing capacity of the adobe wall.

Regarding the deformation capacity of the adobe wall, the maximum horizontal displacements reached in the positive and negative orientations of lateral loading were respectively equal to 31.81 and 34.50 mm (Fig. 15). The height of the horizontal loading line from the base of the piers was $y_0 = 3{,}055$ mm, so the maximum drift experienced by the adobe wall was equal to 1.04% and 1.13% in the positive and negative orientations of lateral loading, respectively. Such a drift capacity cannot be compared to that of the aforementioned TSM wall. Indeed, the latter was

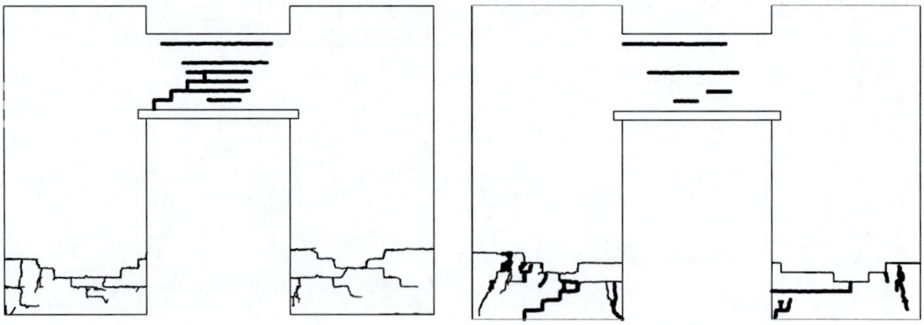

(a) Front (b) Back

Fig. 14 Crack patterns observed on adobe masonry wall with an opening

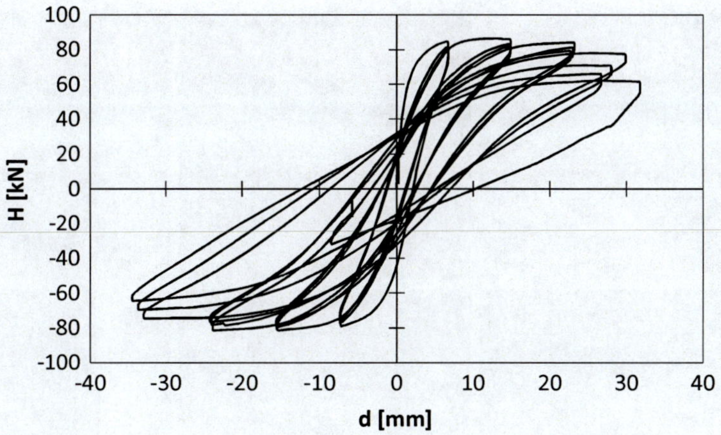

Fig. 15 Force–displacement curve of the adobe wall with a single opening

monotonically tested up to a displacement of 28 mm (corresponding to a maximum drift of 0.89%) that however produced only diagonal shear cracking in the spandrel without causing damage to the piers, and hence a near-collapse condition for the TSM wall. Nonetheless, it can be stated that approximately equal drifts imposed on adobe and TSM walls produced different damage states, i.e. near-collapse and moderate damages.

4 Test Performed at University of Aveiro

At the University of Aveiro, in Portugal, a full-scale double-T shaped adobe wall was built and tested under in-plane horizontal cyclic loading of increasing amplitude. This work, which is briefly presented below, is described in greater detail in Silveira et al. (2018).

4.1 Wall Characteristics

The full-scale adobe wall was built in the laboratory with a double-T horizontal cross-section (Fig. 16) and with the following dimensions: height of 3.07 m; mean thickness of 0.29 m; longitudinal wall length of 3.50 m; transverse wall length of 1.70 m. The adobes used in the wall were collected from an existing adobe house, located in Aveiro municipality, and were made with sandy soil and air lime binder. The mortar was formulated in the laboratory with a composition similar to that traditionally used (hydrated lime: sandy earth ratio of 1:3). The foundation of the wall was made with a reinforced concrete pad footing that was fixed to the

Fig. 16 Construction of the full-scale double-T shaped adobe masonry wall, adapted from Silveira et al. (2018)

laboratory reaction floor with prestressed threaded rods. The first layer of adobes was laid with cement mortar to prevent sliding failure at the base.

4.2 Experimental Test

A horizontal cyclic force of increasing amplitude was applied at a height of 2.60 m from the base of the wall, until failure, using a hydraulic actuator with a maximum load capacity of 100 kN (Figs. 17 and 18). To simulate the dead and quasi-permanent live loads that are typical in adobe buildings, a vertical uniformly distributed load (total value of 20 kN) was added to the top of the wall (Fig. 18b).

The deformation of the wall during the test was measured with displacement transducers (Fig. 18c). A seismograph was also placed on the top of the wall to estimate the evolution of its natural frequency during the test (Fig. 18b). Even though the actuator was force controlled, the system used is controlled manually, which allowed using peak displacements to define the test cycles. In the first cycle, the mean test velocity was approximately 0.5 kN/s and, in the following cycles, it was approximately 1 kN/s. The test cycles, in terms of the maximum horizontal drift imposed, are presented in Table 4.

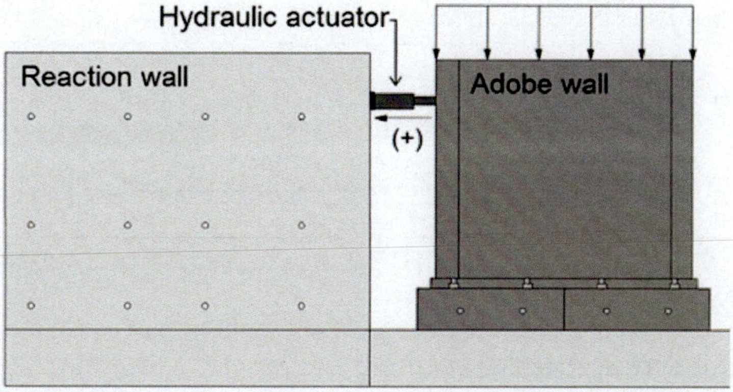

Fig. 17 Test set-up (Silveira et al. 2018)

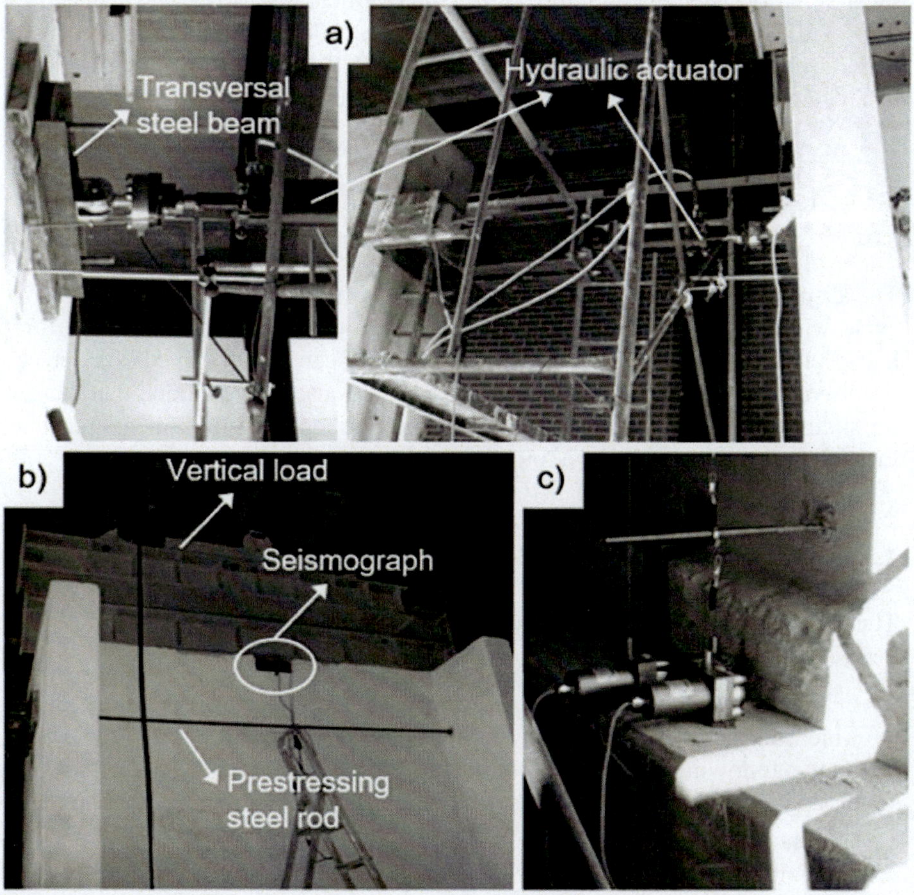

Fig. 18 Test set-up and instrumentation: **a** Force application system. **b** Additional vertical load, seismograph, longitudinal steel rod. **c** Displacement transducers (Silveira et al. 2018)

Table 4 Test cycles (Silveira et al. 2018)

Test cycles					
Maximum drift:	0.01%	0.02%	0.03%	0.05%	• 0.03% at peak stress • 0.16% at 80% of peak stress
No. of repetitions:	2	3	1	1/2	1/2

4.3 Result's Discussion

4.3.1 Stress-Drift Relationship

The shear stress versus horizontal drift curve obtained in the cyclic test is presented in Fig. 19. A maximum lateral force of 58.1 kN, corresponding to a maximum shear stress of 57.3 kPa and 56% of the total vertical dead load, was reached for a drift of 0.03%. The initial tangent shear stiffness, determined as the slope of the stress-drift curve in the initial linear region, is 738 MPa. The wall showed brittle behaviour, with failure occurring suddenly for small values of drift and rotation.

4.3.2 Lateral Displacement Profile

The evolution of the lateral displacement profile of the wall is presented in Fig. 20. The displacement profile evidences an approximately linear response up to the cycle with a maximum drift of 0.03%. From the analysis of the profile for the half-cycle with a maximum drift of 0.05%, it can be observed that cracking occurred at a height between 0.5 and 1.0 m. It can also be concluded that there was no sliding at the base of the wall since the horizontal displacement transducers located near the base displayed negligible displacements.

4.3.3 Dissipated Energy and Equivalent Damping Ratio

The dissipated energy during the test, per test cycle, is presented in Fig. 21. It can be seen from this figure that:

- In the first test cycles (maximum drift of 0.01%), the wall suffered little damage, which translates in low energy dissipation;
- In the following cycles (maximum drifts of 0.02% and 0.03%), the energy dissipation increased steadily as the wall began to suffer damage that propagated slowly;
- In the first half of the last cycle (maximum drift of 0.05%), there was a greater increase in energy dissipation, since this half-cycle corresponds to the formation of the first large diagonal crack.

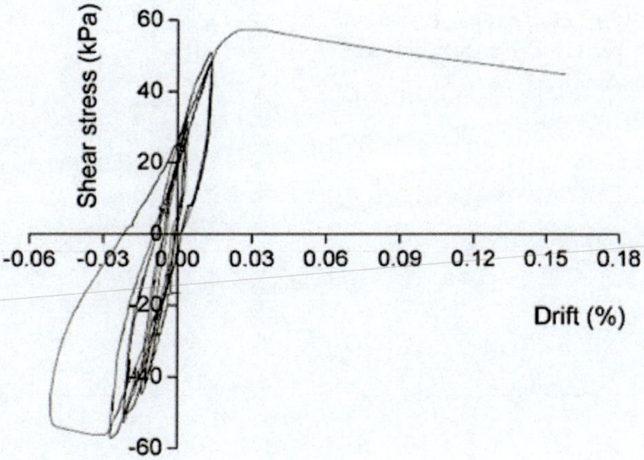

Fig. 19 Shear stress versus horizontal drift curve (Silveira et al. 2018)

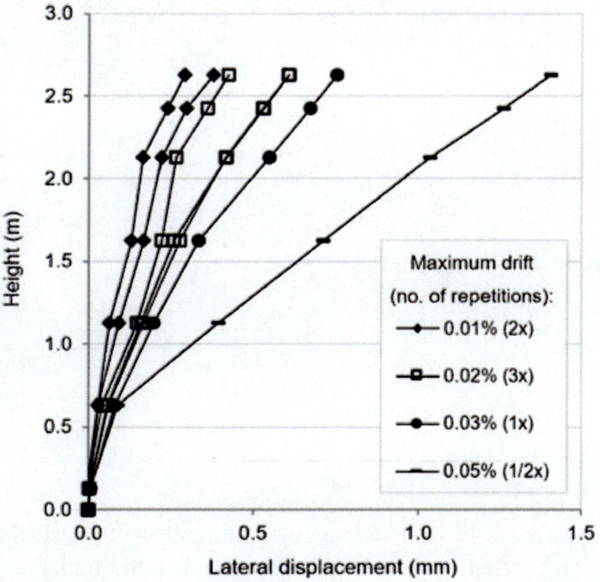

Fig. 20 Evolution of the lateral displacement profile (Silveira et al. 2018)

The equivalent viscous damping ratio was also calculated for the cycles with maximum drift between 0.01% and 0.03%, using the procedure proposed by Varum (2003), based on Jacobsen (1930). For these cycles, an equivalent damping ratio of approximately 15% was calculated.

4.3.4 Natural Frequency

The evolution of the longitudinal natural frequency of the wall during the test is presented in Fig. 22. It can be seen that after the maximum capacity of the wall was reached, there was a greater decrease in frequency, corresponding to a significant

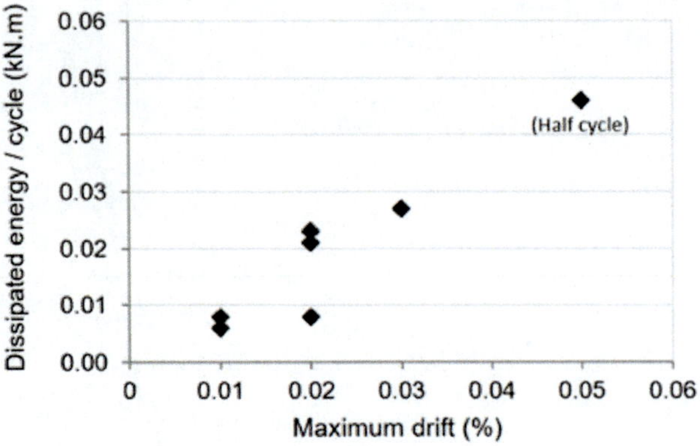

Fig. 21 Dissipated energy per test cycle (Silveira et al. 2018)

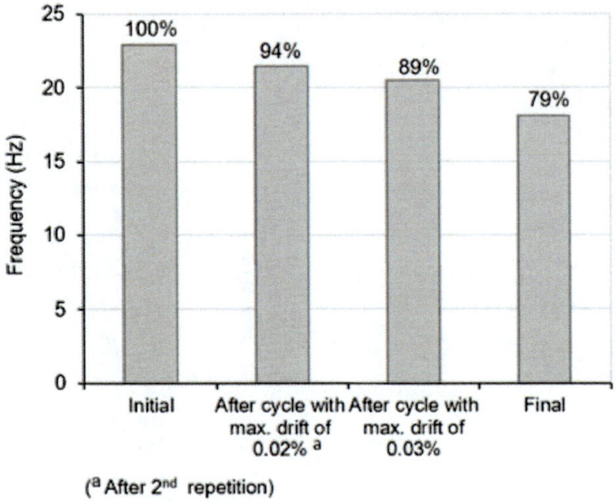

Fig. 22 Evolution of the longitudinal natural frequency (Silveira et al. 2018)

increase in wall damage. At the end of the test, the natural frequency of the wall decreased by 21% when compared to the undamaged wall.

4.3.5 Damage Pattern

Until the last test cycle, the damage that could be observed on the wall was light. The first large diagonal crack was formed during the first half of the last cycle (maximum drift of 0.05%). In the last half-cycle of the test, which corresponds to the onset of failure, another large diagonal crack was formed suddenly in the opposite direction, thus creating the X-shaped pattern that is typical in masonry walls subjected to in-plane cyclic loads (Fig. 23). The cracks in the wall followed mainly along mortar joints, forming a stepped pattern.

Fig. 23 Damage on the wall at the end of the test (Silveira et al. 2018)

5 Conclusions

In this chapter, some pseudo-static in-plane tests developed in three universities have been shown. Those are important to compute the lateral force vs top lateral displacement curve of different adobe wall typologies and with them to establish a simplified in-plane capacity curve. Also, some material and dynamic properties of the masonry, as elasticity module and period of vibration, have been commented. With the results and conclusions of these tests, some numerical modelling could be performed to evaluate the structural behaviour (i.e. damage pattern, load capacity, etc.) of different adobe walls (e.g. variation of geometry) and to establish some vulnerability studies.

Although the geometrical difference in all tests, the structural behaviour of all the walls were quite similar. The results obtained point to very low deformation capacity and brittle failure of adobe walls, evidencing the need for adequate retrofit solutions for these constructions.

References

Augenti N, Parisi F, Prota A, Manfredi G (2011) In-plane lateral response of a full-scale masonry sub-assemblage with and without an inorganic matrix-grid strengthening system. ASCE J Comp Constr 15(4):578–590

Blondet M, Madueño I, Torrealva D, Villa-García G, Ginocchio F (2005) Using industrial materials for the construction of safe adobe houses in seismic areas. In: Proceedings of conference on earth build 2005, Sydney, Australia

Blondet M, Vargas J, Villa-García G (2008) Reparación de grietas en construcciones históricas de tierra en áreas sísmicas. Parte IV: aplicación y desarrollo de nuevos métodos de inyección. Internal report, Division of Civil Engineering, Pontificia Universidad Católica del Perú (in Spanish)

Calvi GM (1999) A displacement-based approach for vulnerability evaluation of classes of buildings. J Earthquake Eng 3(3):411–438
Jacobsen LS (1930) Steady forced vibrations as influenced by damping. ASME Trans 52(1):169–181
Magenes G, Calvi GM (1997) In-plane seismic response of brick masonry walls. Earthquake Eng Struct Dynam 26:1091–1112
Parisi F, Augenti N, Prota A (2014) Implications of the spandrel type on the lateral behavior of unreinforced masonry walls. Earthquake Eng Struct Dyn 43(12):1867–1887
Parisi F, Asprone D, Fenu L, Prota A (2015) Experimental characterization of Italian composite adobe bricks reinforced with straw fibers. Comp Struct 122:300–307
Paulay T, Priestley MJN (1992) Seismic design of reinforced concrete and masonry buildings. Wiley, Hoboken
Silveira D, Varum H, Costa A, Pereira H, Sarchi L, Monteiro R (2018) Seismic behavior of two Portuguese adobe buildings: part I—in-plane cyclic testing of a full-scale adobe wall. Int J Archit Heritage 12(6):922–935
Tarque N, Camata G, Varum H, Spacone E, Blondet M (2014) Numerical simulation of an adobe wall under in-plane loading. Earthquakes Struct 6(6):627–646
Varum H (2003) Seismic assessment, strengthening and repair of existing buildings. PhD thesis, University of Aveiro, Aveiro

Shaking Table Testing of Adobe Masonry Structures

Marcial Blondet, Nicola Tarque, Francisco Ginocchio, and Gladys Villa-García

Abstract This chapter presents the seismic simulation systems (shaking tables) as essential resources for experimental research on adobe masonry structures. An overview of selected relevant shaking tables existing in laboratories around the world gives an idea of the broad testing scope possible with this type of equipment. Its use in testing of adobe masonry structures goes back to 1986 at the University of California at Berkeley and to 1988 at the Pontifical Catholic University of Peru (PUCP). Shaking table testing has been considered the experimental technique that may closest reproduce the real behaviour of an adobe structure during earthquakes. After the description of a typical shaking table test procedure, based on more than thirty years of PUCP experience, some seismic simulation tests on adobe masonry structures performed in different countries are briefly described. Each case presents the basic characteristics of the shaking table and some qualitative results obtained to understand the seismic behaviour of adobe masonry structures.

Keywords Shaking table tests · Adobe masonry · Seismic strengthening

M. Blondet (✉) · N. Tarque · F. Ginocchio · G. Villa-García
Department of Engineering, Pontificia Universidad Católica del Perú,
Av. Universitaria 1801, San Miguel, Lima, Peru
e-mail: mblondet@pucp.edu.pe

N. Tarque
e-mail: sntarque@pucp.edu.pe

F. Ginocchio
e-mail: fginocc@pucp.edu.pe

G. Villa-García
e-mail: gvillag@pucp.edu.pe

H. Varum et al. (eds.), *Structural Characterization and Seismic Retrofitting of Adobe Constructions*, Building Pathology and Rehabilitation 20,
https://doi.org/10.1007/978-3-030-74737-4_6

1 Introduction

Adobe masonry has very low tensile strength compared with that of other engineering materials such as clay brick or concrete block masonry. During seismic movements, it loses strength and stiffness after few oscillation cycles and its mode of failure is brittle. The consequence of medium to high-intensity earthquakes on adobe masonry dwellings is usually considerable damage and in many cases life loss. The common failure modes of adobe walls are overturning due to out-of-plane forces and diagonal cracking due to in-plane forces.

Monotonic or cyclic quasi-static tests are not adequate to obtain realistic information on the seismic response of earthen constructions, since they do not reproduce the effect of accelerations and cyclic dynamic loading. Only dynamic tests on full-scale adobe structures using seismic simulators (shaking tables) provide a reliable way to understand seismic behaviour because they reproduce damage modes similar to those observed in real earthquakes. Furthermore, shaking table testing is especially important to evaluate the adequacy of proposed seismic reinforcement systems for adobe structures. In these cases, tests of unreinforced specimens provide a baseline to assess response improvements.

2 Shaking Tables and Test Set up

The ground shakes in six directions during earthquakes (three translations and three rotations). Therefore, the ideal shaking table would need to reproduce accurately seismic motion recorded in six degrees of freedom. Small tables are available with six degrees of freedom (Fig. 1). Examples are the Hexapod Mast 100 Hz of the German company Inova (2019), and the hydraulic multi-axis Simulation Table of Moog Company (Moog 2019) from the United States.

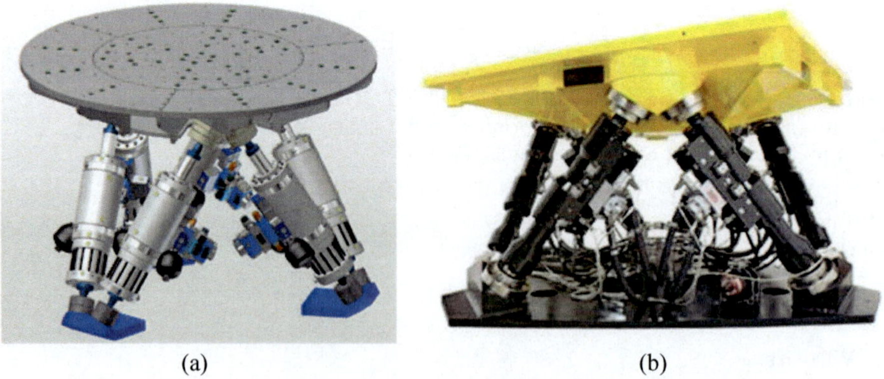

(a) (b)

Fig. 1 Small shaking tables: **a** Inova (2019); **b** Moog (2019)

An example of a large size seismic simulator is the shaking table Azalée at the Laboratory of Seismic Mechanics Studies (EMSI) of the Alternative Energies and Atomic Energy Commission (CEA) in France (CEA 2019). This table, with 6 × 6 m and a mass of 25 t, is the largest in Europe in service since 1990. It allows for the testing of up to 100 t specimens up to 12 m high (Fig. 2). It has eight actuators with a maximum dynamic capacity of 1000 kN each, to provide the triaxial movement with six degrees of freedom, maximum horizontal displacements of ±125 mm (in two directions), and vertical displacements of ±100 mm, with excitation frequencies from 0 to 100 Hz. The maximum acceleration is 1 g when loaded to its maximum capacity.

In Japan, the Hyogo Earthquake Engineering Research Center of the National Research Institute for Earth Science and Disaster Resilience (NIED) has a 20 × 15 m shaking table with a capacity to test full-scale specimens weighing up to 12,000 kN (Sato and Inoue 2004). It has 24 actuators, 10 horizontal (five in each direction), and 14 vertical. The maximum displacements are ±1000 mm (horizontal) and ±700 mm (vertical), with maximum accelerations of 0.9 g (each horizontal) and 1.5 g (vertical) at maximum load. This is the largest shaking table in the world. Its construction began in 1999 and was finished in 2005 (Fig. 3).

In the USA the biggest shaking tables are at two different sites of the University of California. The University of California at Berkeley's shaking table is 6 × 6 m and has six degrees of freedom. Four electro-hydraulic actuators in each direction power horizontal shaking and four actuators power vertical shaking. A system of passive actuators provides rocking reduction. The platform floats on an air cushion, which supports gravity loads of the table and the tested specimen. The shaking table by itself weighs 445 kN and can subject structures, weighing 445 kN, to horizontal accelerations of 1.5 g. Figure 4 shows the shaking table during the test of a 3-phase 245-kV disconnect switch.

Due to the size of the support structure, the disconnect switch was installed on special rigid cantilevered outriggers (in blue-grey color). The shaking platform is surrounded by blue air bladders installed in the gap between the platform and the

Fig. 2 Azalee shaking table (CEA 2019)

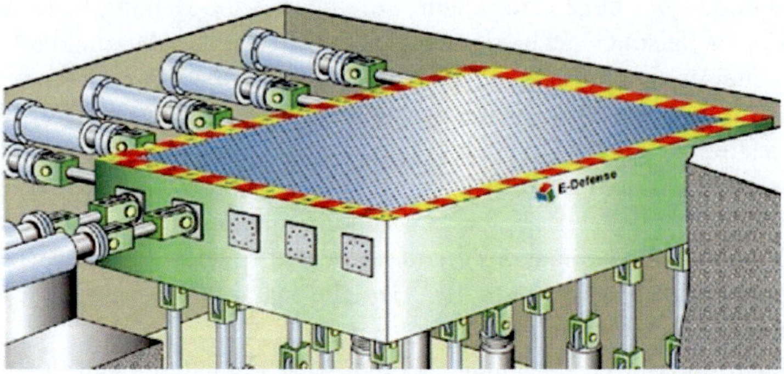

Fig. 3 E-defense shaking table at NIED (Sato and Inoue 2004)

Fig. 4 Testing at UCB shaking table (Takhirov et al. 2017)

reaction mass of the shaking table. Figure 4 shows them pressurized so the weight of the table and the test specimen is supported by the air pressure under the platform. The section of the roof right above the shaking table platform has two sliding roof sections that open for the testing of very tall structures (Shakhzod 2014; Shakhzod et al. 2017).

The Large High-Performance Outdoor Shake Table (LHPOST) at the Natural Hazards Engineering Research Infrastructure (NHERI) of the University of California San Diego (UCSD) (NHERI 2019) is a 7.6 × 12.2 m seismic simulator with an estimated weight of 1440 kN (Fig. 5). The maximum displacement is ±750 mm, with maximum accelerations of 4.2 g (unloaded) and 1.2 g (with 4000 kN maximum payload). It has two servo-controlled dynamic horizontal actuators, with a combined force capacity of 6800 kN and six vertical actuators. The frequency bandwidth ranges from 0 to 33 Hz. This seismic simulation system is the biggest in America.

The kind of equipment described here is extremely expensive and requires highly specialized maintenance. Therefore, they are out of reach for most universities that have smaller shaking tables of only one or two degrees of freedom.

Control by displacement was typical of seismic simulation systems. Double integration and digital filtering of a recorded acceleration generated the displacement control signal. In many cases, the command waveform was a single record and the specified platform motion peak displacement (PGD) or span regulated the shaking intensity. Most modern laboratories have seismic simulation systems with acceleration, velocity, and displacement control. MTS (Measure Test Simulate) provides such advanced control systems (MTS 2020).

The main requirements of a shaking table test program are that the simulated seismic input has approximately the same time envelope and frequency content as the reproduced real earthquake, and that the damage induced in the test specimen is similar to that observed in the field.

The shaking table at the Earthquake-Resistant Structures Laboratory of the Pontifical Catholic University of Per (PUCP) is an example to illustrate the development of a typical seismic simulation test. It was built in 1981 through a Peruvian-Dutch cooperation program and a financial contribution from the Peruvian

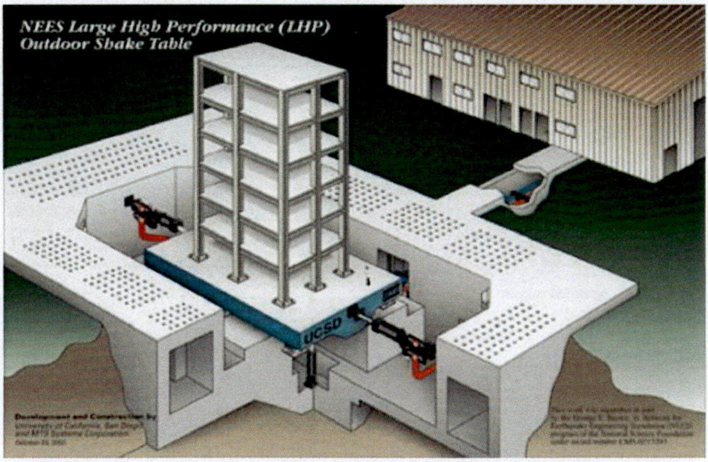

Fig. 5 Shaking table at the University of California San Diego (NHERI 2019)

government. Engineers from the Delft University of Technology designed the seismic simulator.

The seismic simulator consists of a 4.4 × 4.4 m, 180 kN pre-stressed concrete platform (1) supported by eight plates (2) with flexure joints (3) at each end (Fig. 6). It moves in one horizontal direction only. This system enables the testing of structures up to 160 kN. It can reproduce ground motions of up to ±0.15 m displacement amplitude, maximum velocity of 0.5 m/s, and acceleration up to 1.5 g. *A* ± 500 kN horizontal servohydraulic actuator (4) drives the platform. The operating frequency range is 0–12 Hz. The hydraulic system has accumulator banks (5) required to reach large intensity motions. The design of this unidirectional shaking table was to test one-story full-scale earthen models.

The input to the seismic simulator control system is a digital displacement command signal, usually generated from an accelerogram recorded during a real earthquake and processed to comply with the performance specifications of the equipment. The generation of the displacement command signal usually includes digital filtering, double numerical integration and scaling, and baseline correction. The controller then sends an electric (analog) signal to the servo valve that feeds high-pressure oil to drive the actuator attached to the platform. Sensors connected to the computer via a data acquisition system monitor the motion of the platform and the response of the test structure. Figure 7 shows a schematic of the PUCP seismic simulation control and data acquisition system.

Typical instrumentation consists of displacement transducers (LVDTs) and acceleration transducers (accelerometers) placed on the platform and in specific locations of the tested models to measure both seismic excitation and structural response. The actuator has a load cell to measure the force applied to the platform, which is used to estimate the horizontal force applied to the base of the model during shaking.

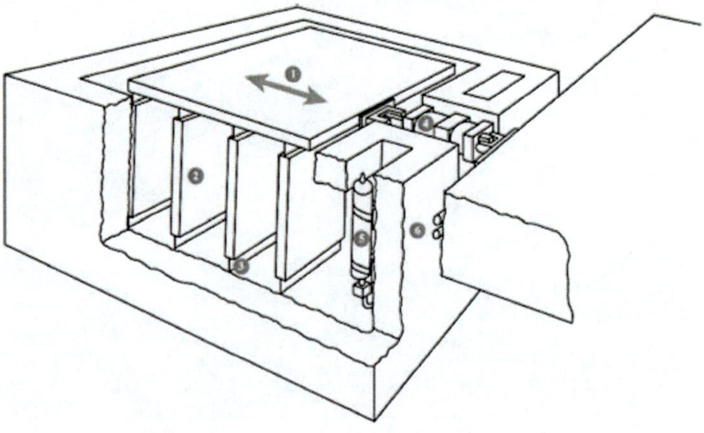

Fig. 6 The seismic simulator at PUCP

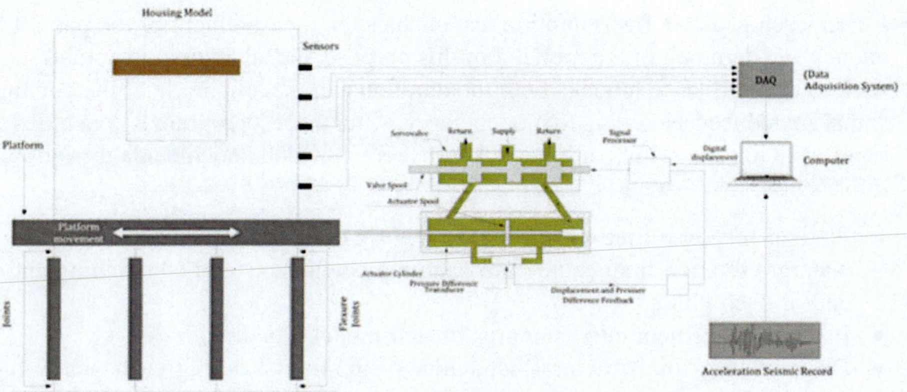

Fig. 7 PUCP seismic simulator control

From the acceleration record captured in Lima by the Geophysical Institute of Peru of the May 31st, 1970 Ancash earthquake (M_W 7.8) (Silgado 1978) derives one of the command signals usually used for earthen models at PUCP, called “mayo70”. With different maximum amplitudes, this signal produces different intensity platform movements. Figure 8 presents typical displacement and acceleration records of the platform motion “mayo70” shaking

Usually, shaking table test campaigns are performed in several steps (also called phases or runs) to represent a sequence of seismic events of increasing intensity. The sequence begins with a small amplitude movement and each consecutive phase increases the motion intensity until partial or total collapse of the test structure occurs.

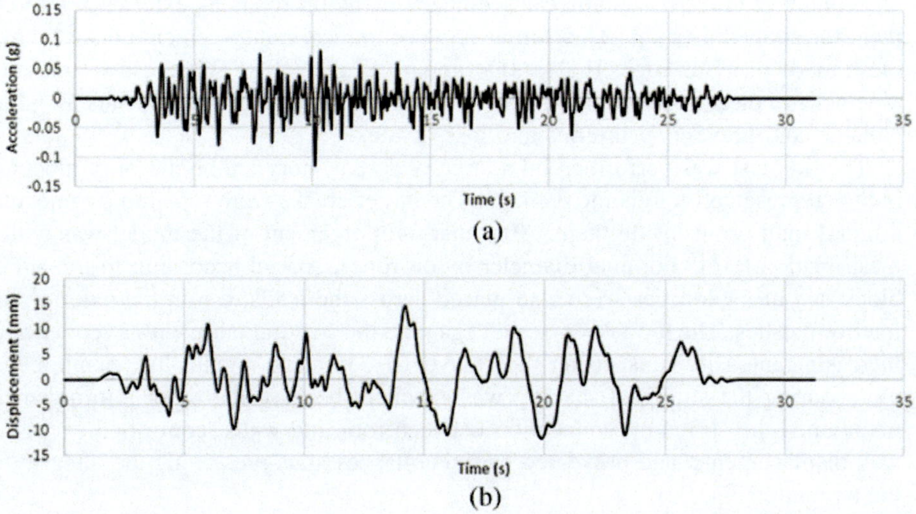

Fig. 8 Platform time histories using “mayo70” command signal: **a** acceleration; **b** displacement

Before each phase, a free vibration test enables the measurement of the natural frequency and damping of the model. For this purpose, the table input is a small square wave. For small-scale models, both the duration and the amplitude of the command signal are reduced according to model theory. After the test program is concluded, the report contains the results of basic data processing, which may include the following information:

- Plots of response time histories recorded by each instrument.
- Natural vibration frequencies and damping coefficients at the beginning and the end of each phase.
- Force-displacement curves and estimated model global stiffness.
- Description of the structural behaviour of the model during each phase, highlighting the evolution of damage and mode of failure.

3 Shaking Table Adobe Research Programs in the World

3.1 Peru

A team of PUCP researchers has been working since the 1970s to find simple and economical ways to provide seismic safety to earthen buildings. Shaking table tests of one-story large-scale housing models evinced the efficacy of each proposed reinforcement system. Some of the reinforcement systems studied included bamboo cane rods, wire mesh covered by cement mortar, polymer geogrid, polymer rope mesh, and nylon rope (Vargas and Ottazzi 1981; Blondet and Esparza 1988; Blondet et al. 2011; Zegarra et al. 1997; Zegarra et al. 2001; Blondet et al. 2016).

Blondet et al. (2016) studied the efficacy of nylon ropes as external reinforcement for adobe buildings. This simple and economical reinforcement consisted of a mesh made of nylon ropes (halyard) wrapping all the walls. This mesh would keep together the pieces of the adobe walls broken during an earthquake, thus avoiding collapse and preventing injuries and loss of life.

The first test was performed on a large-scale one-story adobe masonry model to induce representative seismic damage. The larger cracks were repaired by injecting a liquid mud grout inside them. After that, reinforcement of the model was with a mesh made of 1/4" nominal diameter nylon ropes, spaced according to the adobe block layout. Nylon ties were also placed across the walls to join the exterior and interior meshes. The model was tested again on the shaking table with a sequence of increasing intensity runs, using the "mayo70" command signal. Its seismic behaviour during the strongest shaking was excellent because the mesh reinforcement maintained the structural connection between roof and walls, controlled excessive wall displacements, and prevented partial collapses, thus preserving the integrity of the structure (Fig. 9a).

(a) (b)

Fig. 9 Reinforced adobe models after intense shaking test: **a** 1/4" diameter mesh; **b** 5/32" diameter denser mesh (Blondet et al. 2016)

A proposed simple design method for the rope mesh was validated with the test of an identical one-story large-scale model but reinforced with a thinner 5/32" diameter rope in a denser mesh. The dynamic response of this second specimen was also excellent, as the rope mesh was capable of preventing the overturning of large wall portions during a single strong shaking (Fig. 9b).

These seismic simulation tests on rope-reinforced adobe masonry models revealed that this reinforcing system could be adequate to prevent the collapse of one-story earthen buildings subjected to strong seismic shaking. However, in Peru as in many other Andean countries, many families live in two-story adobe houses. Therefore, the next step was to study the possibility of using the nylon mesh reinforcement technique as seismic reinforcement of two-story adobe houses. As testing of full-scale two-story models could not be on the PUCP's seismic simulator due to its limited mechanical capacity, it was necessary to test reduced-scale models.

Four half-scale two-story adobe building models were built and tested on the shaking table. Two of them had no reinforcement and represented typical Andean two-story adobe houses. The reinforcement of the other two models consisted of a mesh made from 1/8" nominal diameter nylon rope, thus maintaining the scale ratio ($L_{MODEL}/L_{PROTOTYPE} = L_M/L_P = 1/2$) for linear dimensions. Previously the reinforcement for large-scale adobe dwellings, which served as prototypes, had been with 1/4" diameter ropes.

The application of similitude laws, by halving the amplitude of the prototype displacement command signal "mayo70" ($L_M/L_P = 1/2$), and by compressing the time scale by a factor of two ($T_M/T_P = 1/2$) produced the shaking table displacement command signal used to test the scaled two-story adobe models. As expected, the unreinforced models suffered significant damage and were near collapse during moderate shaking. Figure 10a shows the condition of one unreinforced model after shaking with peak displacement amplitude $D_M = 41$ mm and peak acceleration $A_M = 1.27$ g. In a prototype scale these table motions would correspond to ground

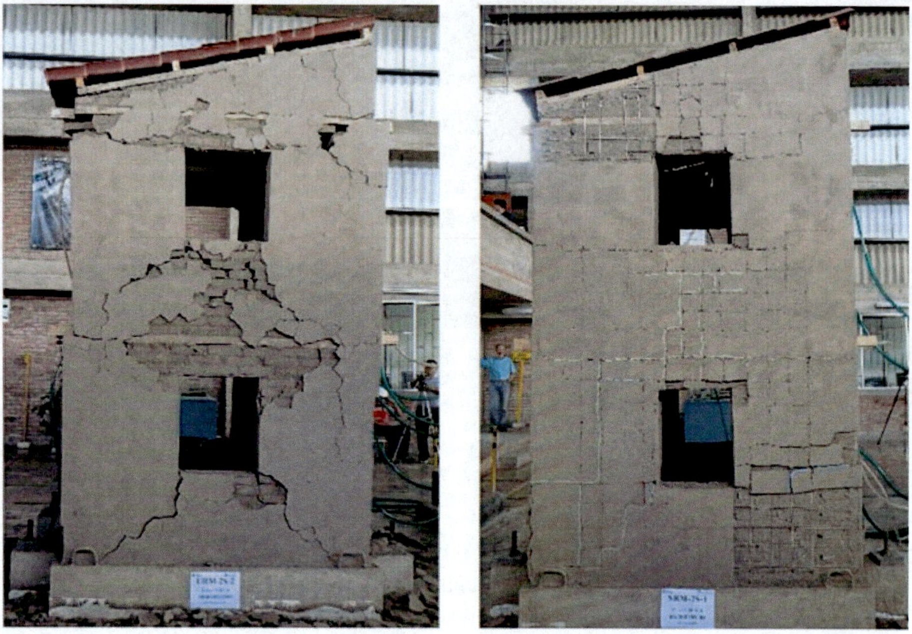

(a) (b)

Fig. 10 Two-story scaled models after shaking table test: **a** unreinforced, close to collapse; **b** reinforced, with moderate damage

motions with peak displacement D_P = 82 mm and peak acceleration A_P = 0.63 g. This is consistent with the field observation that many two-story adobe buildings collapse or are severely damaged after moderate earthquakes.

One of the rope-reinforced half-scale models was subjected to a sequence of motions with increasing intensity. The other model was subjected to the most intense motion that the table could safely provide. In both cases, and similarly to the one-story large-scale prototypes, although adobe walls broke in several large blocks, the rope reinforcement was able to prevent the overturning of these blocks, thus preserving the structural integrity of the models. Figure 10b shows the condition of a reinforced model after test with peak table motion displacement (PGD) D_M = 56 mm and acceleration A_M = 1.67 g (equivalent to D_P = 112 mm and A_P = 0.83 g in prototype scale). The seismic response of the two models was considered to be adequate since the structures were still stable after strong shaking.

3.2 *United States of America*

The first shaking table tests of adobe housing models were at the Earthquake Simulator Laboratory of the University of California at Berkeley (UCB) in 1986, by

Scawthorn (1986) from Dames and Moore. At that time, the seismic simulator had only two degrees of freedom, one horizontal and one vertical (PEER 2018). The test specimen was a 3/4 scale model of a one-room rectangular prototype house with adobe block bearing walls. The plan dimensions were 4.75 × 3.10 m, with 0.30 m thick and 2.12 m high walls. The model had a door and a window in one of the long walls. All the other walls were solid. Wooden beams bearing on the walls supported the flat horizontal roof that had loaded lead weights to simulate a typical heavy roof. The table input signal was derived from the Taft N21E record of the 1952 Kern County earthquake. Four shaking runs were performed, all of them with horizontal input perpendicular to the long walls and with different amplitudes. One of them had both horizontal and vertical input. Low amplitude force vibration before and after the test enabled the estimation of the natural vibration frequency and damping.

The first test with 25.4 mm peak ground displacement (PGD) and 0.08 g peak ground acceleration (PGA) was performed without adding any reinforcement and did not cause visible damage. However, the natural frequency dropped from 15 Hz to about 11 Hz, which indicated some loss of stiffness. The model was then subjected to several additional runs, some of them adding an 11.6 Hz sinusoidal component to the command signal. The final run (PGD = 20 mm, PGA 0.86 g) caused extensive damage at three corners of the model and a natural frequency reduction to 7.5 Hz. The peak acceleration at the corners reached 2.5 g.

The reinforcement of the model consisted of bolting a 0.10 × 3.55 m wooden bond beam to the top of the walls and adding extra weights to the roof. The testing campaign started with three runs until significant diagonal cracking occurred on the longitudinal walls when the table reached its displacement capacity (PGD = 133 mm, PGA = 0.6 g). The two long walls (subjected to out-of-plane motion) and one of the short walls were repaired with mud injection on the major cracks and were reinforced using stapled wire mesh. Several shaking table runs, with different amplitudes and frequency contents were then performed, until the unreinforced wall failed at PGA = 0.7 g. Finally, reparation of all walls was with mud injection and reinforcement was with wire mesh. It was not possible to produce structural failure even after many runs with horizontal motion only (one with PGA = 1.5 g) and horizontal plus vertical motion (one with PGAh = 0.93 g, PGAv = 0.09 g). Figure 11 shows the adobe model on the UCB shaking table. The short walls, subjected to in-plane motions were completely destroyed. However, the wire mesh was able to keep the broken pieces together, thus preserving structural integrity.

This was a pioneer experimental (and analytical) project because it demonstrated that it is possible to avoid the structural collapse of adobe buildings by wrapping all walls with a sufficiently strong mesh, able to keep the broken wall pieces together.

3.3 Mexico

The shaking table of the Structural Dynamics Laboratory at the Engineering Institute of the National Autonomous University of Mexico (UNAM) in the ’80s

Fig. 11 The adobe model after the last shaking table test (Scawthorn 1986)

was 2.4 × 4.5 m with a testing capacity of models up to 150 kN vertical payload. It could reproduce ground motions only in the long direction of the table with a maximum displacement of 25 mm. Five adobe house models on a geometric scale of 1:2.5 were tested (Meli et al. 1980; Hernández et al. 1981) to assess the efficiency of three reinforcement methods to prevent the typical seismic failure of adobe houses (separation of walls in the corners and their subsequent overturning). The acceleration records of El Centro (1940), Managua (1972), and Oaxaca (1973) were the input signals. These earthquakes have high spectral amplitudes for low periods. To comply with dimensional analysis requirements, the signals were corrected and the low frequencies filtered to limit PGD to the actuator capacity.

Three independent models were built, two without reinforcement and one with a reinforced concrete bond beam at the top of the walls. After being tested to a condition very close to total collapse, the two non-reinforced models were strengthened: one with a welded mesh nailed to both faces of the walls and covered by a mortar rendering and the other with horizontal steel rods tied to both faces in the upper part of the walls. The first test had scaled roof trusses and tiles. The other tests had steel channels simply supported on the walls with no restriction to lateral movement and the equivalent mass to increase material density over them.

The roof types and also the table command signals of the first three models were different. After observing that the Oaxaca record produced more severe damage, the remaining tests applied only this record as a command signal, increasing its intensity until reaching model collapse or maximum actuator capacity. To obtain a measure of the amount of damage suffered by the models, after each run a free vibration test was performed to measure the natural period and damping of the model. The conclusion was that all reinforced models showed significantly improved strength, as they were able to withstand without major damage at least twice the maximum seismic intensity supported by the unreinforced models when they were close to collapse. The reinforcement system made of wire mesh was more effective than the use of a concrete bond beam (Fig. 12).

(a) (b)

Fig. 12 Models after shaking table test: **a** reinforced with a concrete bond beam; **b** reinforced with wire mesh (Hernández et al. 1981)

At present, the shaking table at the Structural Dynamics Laboratory is biaxial, has a 4 × 4 m platform, and can test models up to 196 kN payload (Alcocer and Murià 1997). It has four horizontal and four vertical actuators. Its digital technology system can control in real-time 5 degrees of freedom independently or simultaneously: two displacements (longitudinal and vertical) and three rotations. The maximum displacements are ±150 mm horizontally and ±75 mm vertically, with maximum accelerations of 1.2 g horizontally and 2 g vertically. The frequency bandwidth ranges from 0.1 to 50 Hz. Figure 13 shows an adobe housing model reinforced with wire mesh, before testing on this shaking table (Catalán 2013).

3.4 Guatemala

The Seismology Laboratory of the Research Institute of Engineering, Mathematics and Physical Sciences at the Mariano Gálvez University of Guatemala has a 3 × 3 m, bi-axial shaking table with two horizontal actuators, one in each direction. A program to determine the structural response of historical buildings used this facility and had collaboration from MTS Systems Corporation. A first research project was the numerical and experimental study of the Church of San Raymundo in Guatemala, a colonial construction severely damaged by the earthquake of February 4th, 1976 (Torres et al. 2019). A finite element model was developed first, to compare theoretical with experimental results from seismic simulation on a 1:8.75 scale model placed in the transverse direction of the shaking table platform. The model was 3 m long and 1.60 m wide, with 1.70 m height at the front. The height of the vault was 1.20 m. The adobe walls were 0.10 m thick and 0.05 m thick at the vault and dome. The weight, estimated with the mathematical model, was 32 kN. Figure 14 shows the finished model built on the shaking table and the finite element model.

Fig. 13 Adobe model reinforced with wire mesh before shaking table test (Catalán 2013)

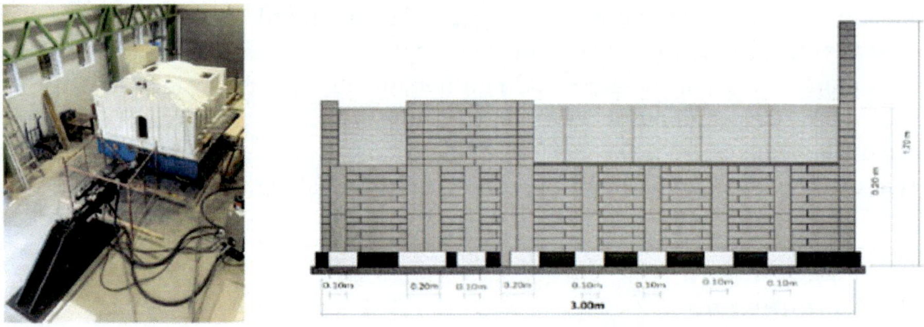

(a) (b)

Fig. 14 Church of San Raymundo finished model **a** front view on shaking table; **b** longitudinal view of analytical model (Torres et al. 2019)

The command signal was derived from the November 7th, 2012 earthquake acceleration record measured in San Marcos, Guatemala. To scale this record the reference was the national standard design spectrum.

The vibration frequencies and modes of the model were analytically estimated from the finite element model. The fundamental frequencies of the first four vibration modes ranged from 5.86 to 8.77 Hz. The model was dynamically tested in its two main directions independently. The experimental failure frequency (6 Hz) was close to the theoretical value for the first vibration mode (5.86 Hz), thus validates the mathematical model analysis. The maximum recorded accelerations were 0.17 g in the walls, 0.28 g in the vault, and 0.47 g in the façade.

The simulated seismic shaking produced very noticeable cracks. For movement in the longitudinal direction, the façade wall (Fig. 15a), tended to overturn due to its

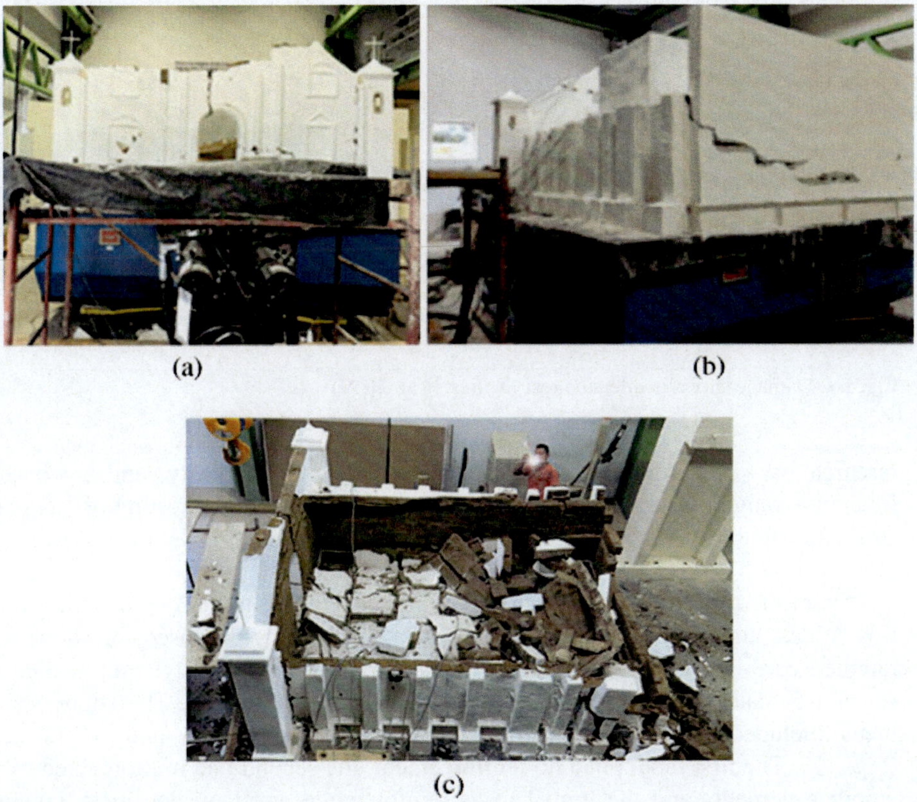

Fig. 15 San Raymundo Church model after testing: **a** damage on façade wall; **b** damage on the rear wall; **c** overall view (Torres et al. 2019)

height and weight, and suffered important damage. For movement in the transverse direction, the façade vibrated independently from the rest of the structure producing bending cracks. The thrust in the longitudinal direction damaged the rear wall (Fig. 15b) due to a lack of out-of-plane restriction. Figure 15c presents the overall view of the model after the test. Figure 16 shows diagonal tension and bending cracks produced in the longitudinal walls. The vault collapsed. The experimental model presented similar damage to that of the Church of San Raymundo during the 1976 earthquake.

3.5 Colombia

The University of Los Andes has two shaking tables. The smaller simulation system (1.05 × 1.05 m), used for academic purposes, has 20 kN vertical bearing capacity and its work frequency ranges from 0 to 20 Hz. The larger table, used mainly for

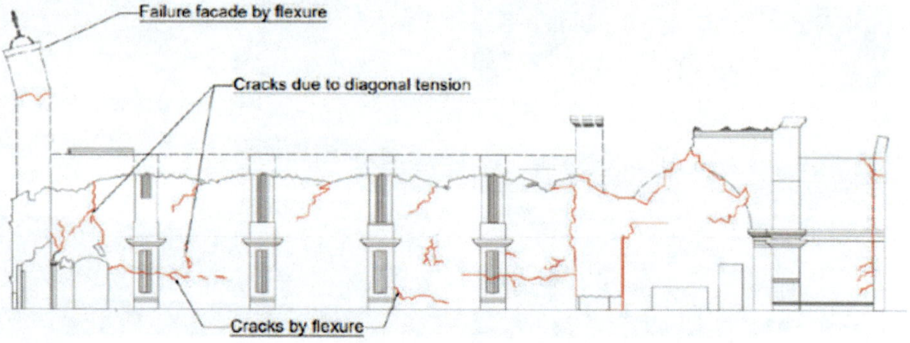

Fig. 16 Damage in walls after the test (Torres et al. 2019)

research, is 4.5 × 4.5 m, has 600 kN vertical bearing capacity, and a working frequency range of 0 to 30 Hz. These characteristics enable the testing of full-scale structures up to three stories according to the specified limits (Clavijo and Ramirez 2011).

One of the milestone projects in Colombia, developed in 2003 by the University Los Andes, aimed to validate two reinforcement systems proposed for adobe and rammed earth structures. The experimental program included dynamic testing of seven 1:5 scaled models in both materials (Yamin et al. 2014). The adobe specimens included three one-floor models 0.6 × 0.6 m in plan and 0.9 m high (Fig. 17). The first model had no reinforcement, the second had reinforcement with wooden elements and the third had as reinforcement steel welded mesh covered with gypsum mortar. To impose all walls with in-plane and out-of-plane base excitation, the orientation of all models was 45° away from the shaking table axis.

The displacement command signal was derived from the acceleration record of the 1995 Tauramena earthquake, registered in El Rosal station. The PGA varied from 0.05 to 2.0 g with 0.05 g increments in each run. Instrumentation consisted of

(a) (b) (c)

Fig. 17 Scaled adobe models built: **a** unreinforced; **b** reinforced with wooden elements; **c** reinforced with steel mesh (Yamin et al. 2014) three accelerometers placed at the base, mezzanine, and roof and two LVDTs at the mezzanine and roof. Measurement of the vibration period and damping for each model was after every run.

The scale of the models did not follow all the similitude laws. One of the reasons was the impossibility to scale the mass of the reduced size structure to compare directly the stress levels with those of the full-scale prototype. Therefore, results were only referred to the qualitative comparison of the models' observed seismic behaviour.

In all cases, the unreinforced models collapsed before the reinforced models (Fig. 18), starting with in-plane and out-of-plane cracking and ending with the overturning of some wall portions. The reinforcement systems modified the global performance of the models subjected to earthquake motion. Although the steel welded mesh increased seismic resistance and delayed collapse, once it reached its maximum strength the models collapsed in an undesirable quasi-brittle way by separation of the reinforced wall areas from the unreinforced wall areas. The best reinforcement system was with the wooden elements because it not only increased seismic capacity but also displacement ductility of the entire structure. At the end of the test runs, this model had damage but did not collapse.

Since then, several other retrofitting solutions have been dynamically tested in similar projects (Yamin et al. 2007; Ruiz et al. 2015; Gómez et al. 2016). An example is the study of the out-of-plane performance of full-scale thick adobe walls (Reyes et al. 2019a; López et al. 2020), which is important because Spanish historic adobe structures usually had very thick walls. The specimens were three L-shaped adobe corner walls 3.0 × 1.8 m in plan, 3.45 m high, and 0.60 m thick. One specimen was unreinforced and the other two had orthogonal timber reinforcement elements. One of these had the vertical elements anchored to the foundation. The three specimens, built on the shaking table, were simultaneously tested. The test protocol consisted of three levels of seismic intensity to correspond roughly to seismic standards levels of damage threshold, limited safety, and design earthquake respectively. The command signal derived from modified versions of the 2008 Quetame earthquake ground motion ($M_L = 5.7$) registered at Bogota Vitelma Station, and the ground acceleration series recorded at the San Francisco

(a) (b) (c)

Fig. 18 Scaled adobe models after testing: **a** unreinforced; **b** reinforced with wooden elements; **c** reinforced with steel mesh (Yamin et al. 2014)

Fig. 19 Final condition of specimens after the last run (PGA = 0.72 g) (Reyes et al. 2019a)

International Airport during the 1989 Loma Prieta earthquake ($M_W = 6.9$). The unreinforced wall collapsed during the second intensity (PGA = 0.39 g) run whereas the other two specimens suffered damaged but did not collapse during the last run (PGA = 0.72 g). Figure 19 shows the final state of the three walls after the last run.

Testing results proved the efficiency of the retrofitting technique with timber elements to improve the out-of-plane integrity of the walls and to prevent unforeseen early collapse by delaying initial crack formation and damage propagation.

Another experimental program involved a series of tests on rammed earth. Thick L-shaped walls, tested on the shaking table, enabled the evaluation of the effect of different types of meshes: steel welded wire mesh and two types of synthetic meshes covering the entire surface of the walls (Reyes et al. 2019b). These dynamic tests proved that under low-intensity motions, meshes had little effect on the out-of-plane seismic response of thick earthen walls. However, when subjected to high-intensity motion, their response depended on the type of mesh. The steel welded wire mesh, the stiffest of the three tested mesh types, produced on the wall the smallest displacement and the largest residual drift during the test. On the other hand, the polyester geogrid mesh produced the smallest residual drift and the least overall damage to the thick wall.

Recently, a reinforcement technique consisting of confinement steel strips placed vertically and horizontally on both faces of the wall was tested on a full-scale L-shaped, 3.50 × 0.60 m wall of rammed earth. To evaluate the effectiveness of the technique, two geometrically similar unreinforced specimens of adobe and rammed earth were also tested (Reyes et al. 2020). The position of the specimens on the unidirectional shaking table was such that it produced out-of-plane movement along the longest wing of the wall. Figure 20 shows the final state of the unreinforced and reinforced specimens. As seen, while the unreinforced wall was unable

(a) (b)

Fig. 20 Final condition of walls: **a** after PGA 0.42 g; **b** after PGA 0.74 g (Reyes et al. 2020)

to resist the complete testing sequence, the reinforcement placed on the wall increased its capacity to withstand out-of-plane activity, minimizing damage and preventing wall collapse.

The Civil Engineering Testing Laboratory of the Pontifical Javeriana University has a unidirectional shaking table of 1.5 × 1.5 m driven by an MTS hydraulic actuator which can generate accelerations of up to 5 g. Its testing capacity is 15 kN with a maximum displacement of ±125 mm (López et al. 2020). The reinforcement for earthen structures was the combination of external wooden elements in both directions and sides of the walls together with an upper concrete beam to create a rigid diaphragm. The tested models were two 1:6 reduced scale, one unreinforced and one fully reinforced. Plan dimensions were 1 × 1 m, the height was 0.57 m and wall thickness was 0.08 m.

The signal was derived from the 1995 Tauramena earthquake. The orientation of the models was 45° from the table axis to impose in-plane and out-of-plane actions on both walls. As expected, the unreinforced model was severely cracked and collapsed before the end of the table shaking. The reinforced model had little damage and around 65 to 80% less wall lateral displacement than the unreinforced model. Figure 21 shows both models at the end of the test.

3.6 Argentina

The shaking table of the Institute for Earthquake Research of the National University of San Juan (Albarracin et al. 2014) is a metal platform 2.90 m long and 2.10 wide supported by two vertical biarticulated plates (Fig. 22). It has one

Fig. 21 Global behaviour of the models after the test: **a** unreinforced; **b** reinforced (López et al. 2020)

Fig. 22 Shaking table at the University of San Juan (Albarracin et al. 2014)

horizontal degree of freedom and its payload testing capacity is 100 kN. A hydraulic actuator commanded by a closed-loop control system moves the table.

The specimens tested on this shaking table were two housing models. To comply with the size and maximum capacity of the seismic simulation system, the scale of the models was 1:2. The first model, with characteristics and dimensions representative of suburban adobe constructions in San Juan, had no reinforcement to obtain baseline information. The second model was reinforced with welded metallic mesh. Both models had cement mortar plastering.

Before the test, the models were subjected to a frequency sweep with low acceleration (0.02 g) to determine their characteristic frequency. It was 10 Hz for the unreinforced model and 20 Hz for the reinforced model. This frequency produced the highest amplification at the ceiling level and was input for the test.

(a) (b)

Fig. 23 Models at the end of the test: **a** without reinforcement; **b** reinforced (Albarracin et al. 2014)

The testing campaign included two parts. The first part had a harmonic 10 Hz command signal with increasing PGA from 0.02 to 0.10 g. The second part had a linear combination of two harmonics, giving a compound signal with increasing PGA from 0.10 to 0.40 g. The frequency of this second part was the characteristic frequency previously determined for each model.

The first part of the test produced no damage to both models. The second part caused severe damage to the unreinforced model which was about to collapse (Fig. 23a). Its behaviour was fragile with shear wall failure that produced wall separation at the corners. The reinforced model had a different and more ductile failure mechanism. The walls did not separate and 45° cracks appeared towards the end of the test. Energy dissipation was concentrated in the lower part of the model, at the interface between the adobe wall and the base foundation structure. However, the structure remained stable (Fig. 23b).

In summary, the proposed reinforcement was effective to increase seismic capacity and delay collapse, improving the general behaviour of the model.

3.7 Portugal

The Earthquake Engineering and Dynamics Division of the National Laboratory of Civil Engineering (LNEC) has a 4.6 × 5.6 m 3D shaking table (LNEC 2020). It has 392 kN payload capacity, three independent orthogonal axes, and a frequency range of 0 to 40 Hz.

On the framework of the SafEarth: Seismic Protection of Earthen Construction Heritage research project (2016–2019) (SafEarth 2019), a two-story adobe building specimen was tested at the LNEC-3D shake table. The building had two façades,

East and West, two sidewalls, North and South, and a gable roof. The scale of the model was approximately 1:1.15 with 3.05 × 4.25 m plan dimensions. Its height was 3.55 m at the roof eaves and 4.55 m at the roof ridge. The construction of the model was inside the LNEC using materials recovered from an old building demolished in the Aveiro region (Candeias et al. 2020).

The testing program of the model included three tests. The first test was on the original model built, without any reinforcement. The specimen for the second test was the same model enhanced by improving connections between floor and walls using steel angles to connect the ends of every one-in-three timber beams to the East and West walls (façades). Also, threaded steel rods connected the middle of the first and last timber beams to the North and South walls (side walls). After this test, repair of the building included repointing the cracks with the same mortar used in the construction, reinforcing them with external steel ties placed at two levels, first floor and roof eaves, and prestressing the ties to an initial tension level of 9 kN each. This was the specimen for the third test.

The displacement command signals were derived from real (El Centro 1940 earthquake) or artificially generated (Marqués de Pombal or Ferradura) records, with different scaled amplitudes. The first test had four phases, each with a different input signal and increasing scale factors. A combination of the Marquês de Pombal and the Ferradura command signals was used with different scaling factors, together with the signal derived from El Centro 1940 earthquake record as an additional test. The command signals used in the second test were only the Ferradura and the Marques de Pombal records. The third test followed the same sequence as the first test with an additional run of the Marques de Pombal record assuming that the building was on rock ground.

Instrumentation consisted of 48 uniaxial accelerometers to measure both in-plane and out-of-plane wall accelerations. Also, two-wire potentiometers measured relative displacements between the East and West walls (façades), one below the first floor and the other below the roof struts, both halfway through the walls' width.

The damage observed in the three tests included cracks in walls in different parts of the building. Figure 24 shows the most relevant damage observed at the end of the second test and Fig. 25 presents damage produced at the end of the third test. The damage was mostly concentrated on the South sidewall, with extensive cracks propagating from the top corners of the door to the top corners of the building. The East façade wall had vertical cracks on the lintel between the two windows and smaller cracks around the window corners. On the West façade wall, an extended vertical crack propagated from the centre of the window down to foundation level, with smaller cracks on the window corners. On the North sidewall there were no significant cracks. After the third test, there were also some horizontal cracks in the corners of the building near the reinforcing steel ties.

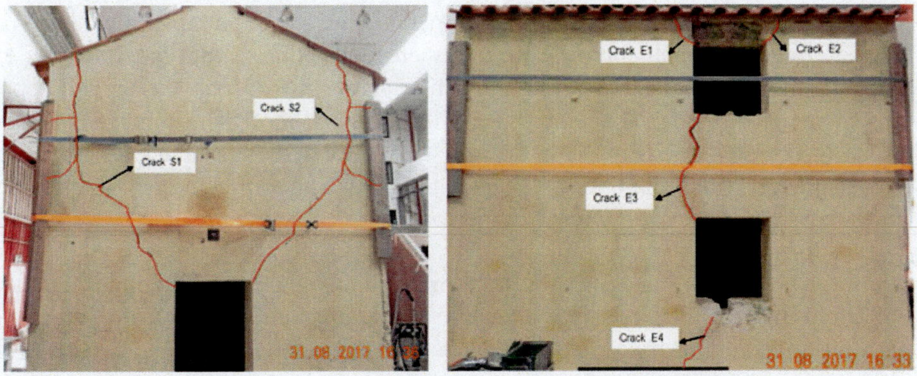

(a) (b)

Fig. 24 After the second test: **a** cracks in the South wall; **b** cracks in the East wall (Candeias et al. 2020)

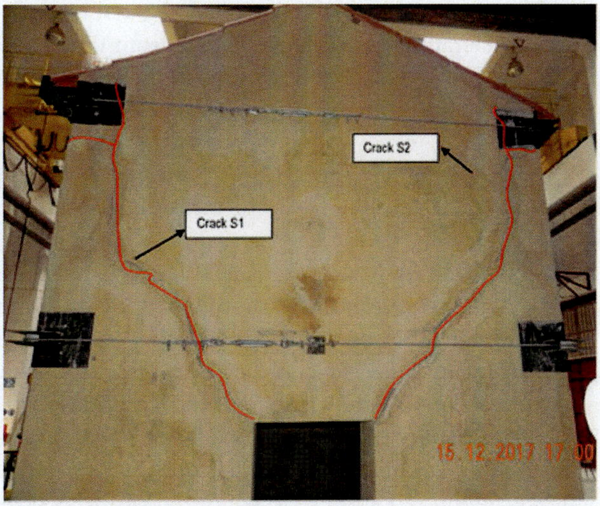

Fig. 25 Cracks in the South wall at the end of the third test (Candeias et al. 2020)

3.8 *Japan*

The shaking table at the Meguro Laboratory of the Institute of Industrial Science (University of Tokyo) is a steel platform 1.5 × 4.5 in plan with 20 kN vertical load capacity and operation frequencies ranging from 0.1 to 50 Hz. The maximum acceleration input, without vertical load, is ±2.0 g in the longitudinal direction and maximum displacement is ±100 mm. This shaking table system is capable of controlling six degrees of freedom.

To understand the dynamic response of adobe masonry houses with and without PP (polypropylene) band mesh retrofitting technique, four unidirectional shaking table tests were performed in 2004 (Sathiparan et al. 2010; Sathiparan et al. 2012). The scale factor, which applied both to dimensions and material properties was 1:4 considering the table size and the allowable loading condition. The models represented a one-story box-like building with a timber roof. The dimensions were 0.933 × 0.933 × 0.720 m with 50 mm thick walls. Mortar thickness between adjacent adobe (unburned brick) layers was 5 mm. The roof consisted of a wooden truss that supported two inclined timber panels.

The first model was unreinforced. The second model was similar to the first but reinforced with PP band meshes 40 mm pitch, prepared with 6 mm-width and 0.32 mm-thick elements. This is a technically feasible and economically affordable retrofitting solution for low earthquake-resistant adobe masonry structures in developing countries. The other two models were similar to the first two except they had surface finish covering applied to both wall sides.

Before the test, a sweep motion of amplitude 0.05 g with frequencies varying from 2 to 35 Hz enabled the identification of the dynamic properties of the models. The natural frequency was 38 Hz for the unreinforced and without surface cover model, and 40 Hz for the other three.

The command signal included sinusoidal motions of frequencies from 2 to 35 Hz and acceleration amplitudes ranging from 0.05 to 1.4 g. The general loading trend was from high frequency to low frequency and from lower amplitude to higher acceleration amplitude, from 0.1 up to 0.8 g. For larger amplitudes, from 1.0 to 1.4 g, the frequency was 5 Hz until the end of the runs.

Instrumentation consisted of twenty-four accelerometers on the house and roof of the model and five lasers to measure displacements.

The shaking table tests showed that scaled adobe housing models with PP-band mesh retrofitting could withstand larger and repeatable shaking as compared to those without retrofitting. The application of surface finishing to the reinforced model improved bond connection between the PP-band mesh and the adobe walls since surface plaster stayed on the walls. The tested reinforcement technique can enhance safety to both existing and new masonry. Therefore, the PP-band retrofitting method can be a good solution to promote safer building construction in developing countries and it can contribute to reducing earthquake damage in the future. Figure 26 shows the condition of the two models without surface covering after the test.

3.9 China

Cast-in-place gypsum-adobe wall structure is a new type of green building for the construction of low-cost rural dwellings in Xinjiang, China (Zhou et al. 2008). To evaluate its seismic performance and to propose improvements, a shaking table test was performed at the Structure and Seismic Testing Laboratory of Xi'an University of Architecture and Technology (Ministry of Education, Key Testing Laboratory).

(a) (b)

Fig. 26 Models without surface finish: **a** non-retrofitted model; **b** retrofitted model (Sathiparan et al. 2012)

The prototype building was a single-floor, two-room structure of 6.90 × 5.10 m plan dimensions, 2.50 m high, and with a 50 mm gypsum layer covering over the walls inside and outside. According to the shaking table size and maximum capacity, the model-scaling factor was 1:4. Therefore, the model dimensions were 1.8 × 1.35 × 0.72 m, external wall thickness was 108 mm, and internal 66 mm. Since the structure was a cast-in-place combination of adobe and gypsum, after framing and placing the adobes, cotton straws inside the joints and continuous connecting wickers laid every two adobe layers, completed the reinforcement. Then, pouring gypsum: water slurry (volume ratio 1:1) completed the compound wall. The roof was made of wood with thermal insulation covering of clay with straw.

Before each test run, the model was hit with a hammer to estimate its dynamic characteristics. The initial natural frequency was 3.7 Hz and decreased (not very fast) until 3.4 Hz, as the input intensity increased. The command signal was derived from the El Centro earthquake record. The test included five phases with increasing intensities.

The instrumentation consisted of accelerometers and displacement transducers. Figure 27 shows the sensor arrangement of the model before testing.

The PGA in each phase was: 0.04 g (moderate intensity 7), 0.08 g (moderate intensity 8), 0.13 g (moderate intensity 8.5), 0.17 g (equal to moderate intensity 9 and rare intensity 8), and 0.24 g (rare intensity 9).

The testing results showed that the structure was able to resist the motions and showed few wall cracks. When peak acceleration of the shaking table was 0.24 g, equal to rare intensity 9, there was a large separation between the model base and the concrete foundation, more cracks developed in walls, but the main structure did not collapse. Since only the compound wall model was tested, it is now possible to have a baseline (traditional construction system) to compare these results.

Fig. 27 Instrumented model before testing (Zhou et al. 2008)

3.10 Australia

A series of dynamic tests on adobe wall specimens was performed at the University of Technology, Sydney (UTS), Australia (Dowling et al. 2005). Its unidirectional shaking table measures 3 × 3 m and can carry a maximum payload of 100kN. It operates in a frequency range of 0.1 to 50 Hz. The maximum acceleration input varies from 2.5 g (unloaded) to 0.9 g (at full load) and the maximum displacement is ±100 mm.

The testing program included five U-shaped adobe walls at a 1:2 scale. The first wall had no reinforcement, the second was also without reinforcement but had corner buttresses. The reinforcement of the other three specimens was with internal horizontal chicken wire mesh (every three courses), a timber ring beam, and vertical bamboo rods (external and in one case internal). For each specimen, the length of the long wall was 1.80 m, each wing wall was 1.20 m long and the height was 1.20 m. The adobe blocks, laid in a stretcher bond, gave 0.15 m wall thickness. A downward restraining force, applied to the tops of the wing walls, prevented overturning of the specimen. The position of the wall on the table was such that the long wall was subjected to out-of-plane action.

The displacement command signal was derived from the January 13, 2001, El Salvador earthquake (Mw 7.7). A modal hammer test helped identify the natural frequency of each specimen. Scaling of the input time histories was to induce damaging resonance conditions. Also scaling of the shaking table displacement was to produce ground motion simulations of varying intensities. To study the behaviour and performance of the structures at different load levels, the testing sequence included a series of runs with increasing displacement intensities, from 20 to 125% for the time-scaled input. Figure 28 presents the unreinforced specimen after the 75% displacement test and the specimen reinforced with external vertical bamboo after the 100% displacement test.

The shaking table test results indicate that the reinforcement system consisting of external vertical and/or horizontal bamboo cane rods, internal horizontal chicken

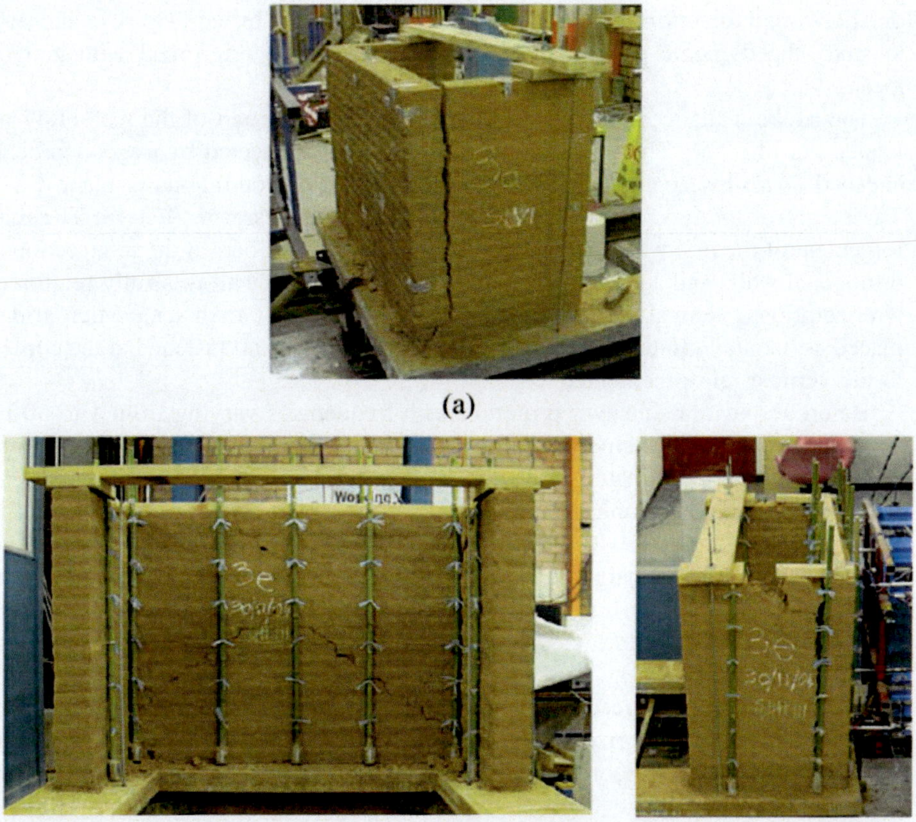

(a)

(b)

Fig. 28 Specimens after dynamic testing: **a** unreinforced; **b** reinforced (Dowling et al. 2005)

wire mesh, and a ring beam, is effectively retarding initial cracking. It also produces significant improvement in the earthquake response, delaying major structural damage and ultimate collapse. However, there is some uncertainty related to the structural performance of internal vertical reinforcement, because significant cracking occurred at lower intensities than expected. Test results also reveal that damage on U-shaped adobe wall panels showed classic failure patterns consistent with those observed on real structures subjected to real earthquakes.

3.11 New Zealand

The shaking table at the University of Auckland measures 1 × 1.5 m. Its maximum payload is 10 kN and it operates in a frequency range of 1 to 50 Hz. The maximum acceleration input with load is 1.68 g and the maximum displacement is 100 mm in one horizontal direction. Two U-shaped adobe walls were tested (Tipler et al. 2010) to study the dynamic performance of these walls when reinforced with geogrid mesh.

The adobe walls, constructed at 1:3 scale had the long part of the wall (1.47 m) loaded out of plane. The wing walls (0.65 m each), connected by a wooden beam and tied down by steel rods, simulated the restraint of continuous in-plane walls. The height of the model was 0.8 m and the wall thickness was 0.10 m. The reinforcement of the first wall consisted of vertical 4 mm steel bars at the intersection of orthogonal walls and a structural bond beam. The second wall was fully reinforced with additional vertical steel and a polysynthetic geogrid (small-scale Microgrid®) placed horizontally in the mortar layer every third course and fastened at the corners to the vertical reinforcement.

Before the test, a sine sweep motion with frequencies varying from 1 to 50 Hz applied to the models helped in identifying their dynamic properties. The natural frequency was approximately the same, 12 Hz, for both models.

The acceleration command signals were derived from three earthquake records: El Centro, California (1940); Northridge, California (1994); and, Llolleo, Chile (1985). The testing campaign included running the records at the serviceability limit state (SLS) and ultimate limit state (ULS), scaling the PGA according to the procedure set out in *NZS 1170.5 Structural Design Actions: Part 5—Earthquake Actions* (Standards 2004).

In the partially reinforced wall, a large vertical crack formed at mid-span due to flexural stresses in the horizontal span, which propagated into a diagonal crack after the Llolleo earthquake record run. The fully reinforced wall did not collapse during the test after running the three earthquake records. Then, additional sine sweep motions, with frequencies ranging from 2 to 16 Hz and with maximum acceleration up to 1.6 g, were successively applied to damage the wall.

The qualitative analysis showed that geogrid reinforcement improved the resistance of the adobe walls to reach collapse. In both walls cracking occurred through the adobe bricks as well as the mortar. In a full-scale wall cracking usually runs mainly through the mortar layer. This scale effect occurs because smaller bricks have a greater surface area to volume ratio than larger bricks and thus, the small adobe block masonry tends to act monolithically. The failure mechanism in both walls resulted from a combination of horizontal and vertical flexural cracking.

4 Concluding Remarks

Earthen constructions such as adobe buildings have shown undesirable and harmful performance during earthquakes. A shaking table is therefore a piece of equipment that is essential to understand the response of traditional adobe construction and to evaluate the effectiveness of seismic reinforcement techniques.

An ideal seismic simulator would reproduce movement with six degrees of freedom (three displacements and three rotations). Such a facility, however, would require high implementation and maintenance costs. Therefore, research institutions and universities have acquired more accessible shaking tables with one or two degrees of freedom. Together with the application of model scaling theories, these facilities enable testing that can reasonably reproduce ground shaking and field damage patterns on adobe structures. The use of reduced-scale models, however, requires special caution because adobe masonry has an extremely complex seismic response which is difficult to model experimentally and analytically.

Over time, this type of experimental research will enable the development of successful reinforcement systems that could prevent extensive damage or collapse of adobe structures. Shaking table test results could also provide a technical basis to adobe construction codes or building policies in seismic countries (SENCICO 2000; SNZ 1998).

References

Albarracin O, Saldivar M, Garino Libardi L, Navarta G (2014) Reforzamiento de Estructuras de Adobe con Mallas Metálicas. In: Seminario Iberoamericano de Arquitectura y Construcción con Tierra SIACOT 2014. Red Proterra. San Salvador, El Salvador

Alcocer SM, Murià D (1997) La nueva mesa vibradora del Instituto de Ingeniería de la Universidad Nacional Autónoma de México. Informe final Proyecto 6539 preparado para el Consejo Nacional de Ciencia y Tecnología. Mexico

Blondet M, Esparza C (1988) Analysis of shaking table-structure interaction effects during seismic simulation tests. Earthq Eng Struct Dynam 16:473–490

Blondet M, Vargas J, Tarque N, Iwaki C (2011) Seismic resistant earthen construction: the contemporary experience at the Pontificia Universidad Católica del Perú. J Informes de la Construcción 63(523):41–50

Blondet M, Vargas J, Tarque N, Sosa C, Soto J, Sarmiento J (2016) Seismic protection of earthen vernacular and historical constructions. In: Balen V, Verstrynge (eds) SAHC—structural analysis historical constructions—anamnesis, diagnosis, therapy, controls. CRC Press

Candeias P, Carvalho A, Correia A, Campos Costa A (2020) Characterization of the seismic behaviour of adobe and rammed earth structures. Experimental seismic tests: adobe structures. Project SafEarth—Seismic protection of earthen construction heritage, POCI-01-0145-FEDER-016737 and PTDC/ECM-EST/2777/2014. Report DE/NESDE, LNEC

Catalán P (2013) Tesis Comportamiento sísmico de la vivienda de adobe basado en pruebas en mesa vibradora de dos modelos a escala. Master thesis. Instituto de Ingeniería. Universidad Nacional Autónoma de México UNAM. Mexico

CEA (2019) Laboratory of seismic mechanic studies (EMSI). http://www-tamaris.cea.fr/html/en/tests/azalee.php. Accessed on Aug 2019

Clavijo J, Ramirez L (2011) Diseño, modelamiento y simulación de una mesa sísmica unidireccional hidráulica. Undergraduate thesis. Facultad de Ingenierías Físico Mecánicas, Escuela de Ingeniería Mecánica, Universidad Industrial de Santander. Bucaramanga, Colombia

Dowling D, Samali B, Li J (2005) An improved means of reinforcing adobe walls—external vertical reinforcement. In: SismoAdobe2005. Pontificia Universidad Católica del Perú. Lima, Peru

Gómez V, López C, Ruiz D (2016) Rehabilitación sísmica de edificaciones históricas en tapia pisada: estudio de caso de capillas doctrineras reforzadas con malla de acero y madera de confinamiento. Informes de la Construcción 68(541):e140

Hernández O, Meli R, Padilla M, Valencia E (1981) Refuerzo de la Vivienda Económica en Zonas Sísmicas. Series del Instituto de Ingeniería 441. Universidad Nacional Autónoma de México

Inova (2019) Institutional Web Page. https://www.inova-gmbh.com/applications/hexapod-test-rigs/, Germany

LNEC Laboratório Nacional de Engenharia Civil (2020) Institutional Web Page. http://www.lnec.pt/en/research/research-infrastructures/shaking-tables/, Portugal. Accessed on Aug 2020

López C, Ruiz D, Jerez S, Aguilar S, Torres J, Alvarado Y (2020) Comportamiento sísmico de edificaciones de tapia pisada reforzadas con marcos de madera y viga de coronación de concreto. Informes de la Construcción 72(559):e347

Meli R, Hernández O, Padilla M (1980) Strengthening of adobe houses for seismic actions. In: Seventh world conference on earthquake engineering. Istanbul, Turkey

Moog (2019) Institutional Web Page. http://www.moog.com/products/simulation-tables/hydraulic-simulation-table/, USA

MTS (2020) MTS systems corporation. https://test.mts.com/products/civil-engineering/seismic-simulators. Accessed on Oct 2020

NHERI (2019) Facilities. Shake table specifications. http://nheri.ucsd.edu/facilities/shake-table.shtml, UC San Diego, USA. Accessed on Aug 2019

PEER: Pacific Earthquake Engineering Research Center (2018) UC Berkeley Shaking Table. https://peer.berkeley.edu/laboratories/uc-berkeley-shaking-table, UC Regents, Berkeley

Reyes J, Galvis F, Yamin L, Gonzalez C, Sandoval J, Heresi P (2019a) Out-of-plane shaking table tests of full-scale historic adobe corner walls retrofitted with timber elements. Earthq Eng Struct Dyn 48(8):888–909

Reyes C, Smith-Pardo J, Yamin L, Galvis F, Angel C, Sandoval J, Gonzalez C (2019b) Seismic experimental assessment of steel and synthetic meshes for retrofitting heritage earthen structures. Eng Struct 198:109477

Reyes J, Rincón R, Yamin L, Correal J, Martinez J, Sandoval J, Gonzalez C, Angel C (2020) Seismic retrofitting of existing earthen structures using steel plates. Constr Build Mater 230:117039

Ruiz D, López C, Unigarro S, Domínguez M (2015) Seismic rehabilitation of sixteenth- and seventeenth-century rammed earth–built churches in the Andean highlands: field and laboratory study. J Perform Constructed Facilities 29(6)

SafEarth (2019) Seismic protection of earthen construction heritage (research project, 2016–2019), with reference POCI-01-0145-FEDER-016737 (PTDC/ECM-EST/2777/2014). University of Minho, National Laboratory of Civil Engineering (LNEC), University of Aveiro, and University of Porto (FEUP), Portugal

Sathiparan N, Mayorca P, Meguro K (2010) Experimental study on static and dynamic behaviour of PP-band mesh retrofitted adobe masonry structure. In: 7th International conference on urban earthquake engineering (7CUEE) & 5th international conference on earthquake engineering (5ICEE). Tokyo Institute of Technology. Tokyo, Japan

Sathiparan N, Mayorca P, Meguro K (2012) Shaking table tests on ¼ scale PP-band retrofitted model of low earthquake resistant masonry houses. Earthquake Spectra 28(8):277–299

Sato M, Inoue T (2004) General frame work of research topics utilizing the 3-D full-scale earthquake testing facility. J Japan Assoc Earthq Eng 4(3):448–456

Scawthorn C (1986) Strengthening of low-strength masonry buildings: analytical and shaking table test results. In: Middle East and Mediterranean regional conference on earthen and low-strength masonry buildings in seismic areas. Middle East Technical University. Ankara. Turkey

SENCICO (2020) Norma Técnica de Edificación NTE E.080 Adobe. Reglamento Nacional de Construcciones. Lima, Peru

Shakhzod MT (2014) Seismic test—qualification report of EV-1, 245 kV, 3000 A, 900 kV BIL disconnect switch on SDG&E Structure. PEER-STI client report no. PEER—STI/2014-11. Pacific Earthquake Engineering Research Center, University of California, Berkeley, p 88

Silgado E (1978) Historia de los sismos más notables ocurridos en el Perú (1513–1974). Boletín No. 3. Serie C. Geodinámica e Ingeniería Geológica. Instituto de Geología y Minería (INGEOMIN). Lima, Peru

SNZ (1998) Engineering design of earth buildings. NZS 4297:1998. Wellington: Standards New Zealand

SNZ (2004) NZS1170.5. Structural design actions part 5 earthquake actions—New Zealand. Standards New Zealand, Wellingto

Takhirov S, Blalock F, Stewart J (2017) Experimental evaluation of system level properties of porcelain post insulators based on a large set of full-scale high-voltage insulators. In: 7th international conference on advances in experimental structural engineering, Sept 6–8,Pavia, Italy

Tipler J, Worth M, Morris H, Ma Q (2010) Shake table testing of scaled geogrid-reinforced adobe wall models. In: 2010 New Zealand Society of Earthquake Engineering Conference. Wellington, New Zealand

Torres R, Blanco A, De Paz J, Campo J (2019) Experimental dynamic behaviour of historical buildings at scale model: San Raymundo Church, Guatemala. In: Conference proceedings, structural analysis of historical constructions: an interdisciplinary approach, RILEM Bookseries pp 2501–2509

Vargas J, Ottazzi G (1981) Investigaciones en adobe, Report. Division of Civil Engineering, Pontificia Universidad Católica del Perú, Lima, Peru

Yamin L, Phillips C, Reyes J, Ruiz D (2007) Seismic vulnerability studies, renovation, and reinforcement of houses built with adobe brick and rammed earth. Apuntes: Revista de Estudios Sobre Patrimonio Cultural—J Cultural Heritage Studies 20(2):286–303

Yamin L, Rodríguez Á, Fonseca L, Reyes J, Phillips C (2014) Comportamiento sísmico y alternativas de Rehabilitación de edificaciones en adobe y tapia Pisada con base en modelos a escala reducida Ensayados en mesa vibratoria. Revista De Ingeniería 0(18):175–190.

Zegarra L, Quiun D, San Bartolomé A, Giesecke A (1997) Reinforcement of existing adobe dwellings 2nd part: seismic test of models (in Spanish). In: XI National Congress on Civil Engineering, Trujillo, Peru

Zegarra L, Quiun D, San Bartolomé A, Giesecke A (2001) Behaviour of reinforced adobe houses in Moquegua, Tacna, and Arica during the June 23, 2001 earthquake (in Spanish). In: XIII National Congress on Civil Engineering, Puno, Peru

Zhou T, Hu X, Yu C (2008) Shaking table model test and engineering practice of a new gypsum-adobe walls dwelling in Xinjiang autonomous region, China. Acta Seismological Sinica 21:319–324

Chapter 11
Axially Functionally Graded Pipes Conveying Fluid

11.1 Introduction

Pipelines conveying fluid exist widely in many application fields, particularly in nuclear power plants, chemical plants, aeronautic, oil transportation, water supply, heat exchanger devices, human circulation, etc. The high velocity internal flow may cause severe flow-induced vibration of piping systems, which may further result in leakages, fatigue failures, high noise, fire, and explosions of the pipes (Sadeghi and Karimi-Dona 2011). Extensive investigations have been carried out in the past decades to understand the dynamical behavior of pipes conveying fluid, as described by Païdoussis and Li (1993) and Païdoussis (1998, 2008).

Similar to other structural dynamic problems, the earliest concern of fluid-conveying pipes was the free vibration response (Long 1955). Research reveals that the boundary conditions can affect significantly the natural frequencies of the dynamic systems (Païdoussis and Issid 1974; Païdoussis 1975). Some numerical methods such as homotopy perturbation method (Xu et al. 2010) and precise integration method (Liu and Xuan 2010) were developed to analyze the effect of fluid flow velocity on the natural frequencies. Although there are many studies considering the flow velocity as constant, the flow velocity varies with time for the actual industrial problems. The unsteady flow is usually modeled by the superposition of the steady flow and a time-dependent harmonic component, which may induce the dynamic instability due to parametric resonances (Ginsberg 1973). For instance, Jin and Song (2005) investigated the effect of some physical parameters of the system, such as damping, mean flow velocity, mass ratio, tension, and gravity, on the three regions of parametric resonances of pipes with supported ends conveying pulsating fluid. Panda and Kar (2008) analyzed the nonlinear planar vibration of a hinged–hinged pipe conveying fluid with harmonic flow velocity pulsation in the presence of internal resonance. With the development of material technology and application, new materials exhibiting viscoelastic behaviors such as polymer matrix composites are now widely used for pipes. Zhao et al. (2001) investigated the

C. An et al., *Structural and Thermal Analyses of Deepwater Pipes*,
https://doi.org/10.1007/978-3-030-53540-7_11

dynamic behavior and stability of Maxwell viscoelastic pipes conveying fluid with simply-supported ends. Zhang et al. (2001) presented a viscoelastic finite element approach to the vibration analysis of viscoelastic Timoshenko pipes conveying fluid. Wang et al. (2005) studied the vibration and stability of viscoelastic curved pipes conveying fluid using normalized power series method. Yang et al. (2007) investigated the dynamic stability for the transverse vibrations of pipes conveying fluid using the method of multiple scales. To avoid the failure caused by resonance due to the excitation of external forces, many researchers considered the forced vibrations of pipes conveying fluid. Gulyayev and Tolbatov (2002) carried out the numerical modeling of self-excited vibrations of tubes containing inner flows of non-homogeneous boiling fluid. Seo et al. (2005) presented the finite element method to predict the forced vibration response of a pipe conveying harmonically pulsating fluid. Liang and Wen (2011) studied the forced responses with both an internal resonance and an external periodic excitation of the constant-fluid-conveying pipe by the multidimensional Lindstedt–Poincaré method. In practice, most of the load applied on the industrial pipes is random, therefore, the dynamic response of pipes conveying fluid subjected to random excitation was studied by Zhai et al. (2011) and Zhai et al. (2013). Since the high-temperature environment should be confronted in some industries such as nuclear reactors, space planes and chemical plants, the vibration behaviors of pipes conveying fluid under thermal loads have been studied in recent years (Kadoli and Ganesan 2004; Ganesan and Kadoli 2004; Sheng and Wang 2008; Qian et al. 2009; Hosseini and Fazelzadeh 2011).

Due to the advantages of being able to withstand severe high-temperature gradient while maintaining structural integrity, functionally graded materials (FGMs) have attracted great interest in a broad range of applications including biomechanical, automotive, aerospace, mechanical, civil, nuclear, and naval engineering (Shen 2009; Zhong et al. 2012). As it is known that FGMs are a novel class of composite materials whose composition and/or function is designed to change continuously within the solid. The composites are usually made from a mixture of metals and ceramics to ensure the elastic and toughness properties gradually vary in space, which can prevent delamination and stress concentration in traditional multilayer, laminated composites. For pipes conveying fluid, Sheng and Wang (2008) reported the result of an investigation into the coupled vibration characteristics of fluid-filled functionally graded cylindrical shells, while Hosseini and Fazelzadeh (2011) investigated the thermomechanical stability of functionally graded thin-walled cantilever pipes conveying flow and loading by compressive axial force. Both of the above-mentioned investigations assumed that the material properties vary along the thickness direction of pipes, however, dynamic behaviors of axially functionally graded systems (structures with material graduation through the longitudinal directions) should be also concerned, as reported by Huang and Li (2010), Huang et al. (2013), Shahba et al. (2011), Shahba and Rajasekaran (2012), Simsek et al. (2012), and Alshorbagy et al. (2011). Nevertheless, the literature dealing with the dynamic behavior of fluid-conveying pipes made of axially FGMs is very limited, which motivates the continuing research on this topic.

In this chapter, the dynamic behavior of axially functionally graded (FG) pipes conveying fluid is analytically and numerically investigated on the basis of the generalized integral transform technique (GITT), which has been successfully applied in solving the dynamic response of axially moving beams (An and Su 2011), axially moving orthotropic plates (An and Su 2014a), fluid-conveying pipes (Gu et al. 2013) and pipes conveying gas–liquid two-phase flow (An and Su 2015), the wind-induced vibration on overhead conductors (Matt 2009), the vortex-induced vibration of long flexible cylinders (Gu et al. 2012), and the transverse vibrations of a cantilever beam with an eccentric tip mass in the axial direction (Matt 2013b). From an engineering viewpoint, this chapter generates the reliable reference data on the dynamic behavior of axially FG conveying fluid, while from a mathematical viewpoint, the chapter provides a feasible numerical solution of the variable coefficient partial differential equations governing the phenomenon. The chapter is organized as follows: In the next section, the mathematical formulation of the transverse vibration problem of axially FG pipes conveying fluid is presented. In Sect. 11.3, the hybrid numerical–analytical solution is obtained by carrying out integral transform. Numerical results of proposed method including transverse displacements and their corresponding convergence behavior and verification are presented in Sect. 11.4. A parameter study is then performed to investigate the effects of material distributions and mass ratios on natural frequencies and vibration amplitude of pipes conveying fluid, respectively. Besides, the variation of the dimensionless frequencies with Young's modulus ratio, power exponent and flow velocity are also presented. Finally, the chapter ends in Sect. 11.5 with conclusions and perspectives.

11.2 Mathematical Formulation

We consider a fluid-conveying pipe made of axially functionally graded (FG) material based on Euler–Bernoulli beam theory, as illustrated in Fig. 11.1. If gravity, internal damping, external imposed tension, and pressurization effects are either absent or neglected, the equation of motion of the FG pipe can be derived following

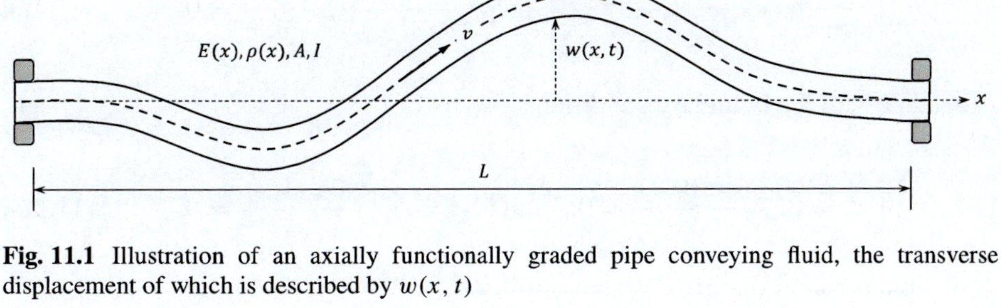

Fig. 11.1 Illustration of an axially functionally graded pipe conveying fluid, the transverse displacement of which is described by $w(x, t)$

the Newtonian derivation by means of decomposing an infinitesimal pipe-fluid element into the pipe element and the fluid element, according to the procedure given by Païdoussis (1998):

$$\frac{\partial^2}{\partial x^2}\left[E(x)I\frac{\partial^2 w}{\partial x^2}\right]+m_f v^2\frac{\partial^2 w}{\partial x^2}+2m_f v\frac{\partial^2 w}{\partial x\partial t}+\left[m_f+\rho(x)A\right]\frac{\partial^2 w}{\partial t^2}=0, \tag{11.1a}$$

subjected to the clamped–clamped boundary conditions

$$w(0,t)=0,\quad \frac{\partial w(0,t)}{\partial x}=0,\quad w(L,t)=0,\quad \frac{\partial w(L,t)}{\partial x}=0, \tag{11.1b–e}$$

where $w(x,t)$ is the transverse displacement, $E(x)I$ is the flexural rigidity of the pipe which depends upon both Young's modulus $E(x)$ and the inertial moment of cross-sectional area I, m_f is the mass of fluid per unit length, v is the steady flow velocity, $\rho(x)A$ is the mass of the pipe per unit length which depends upon mass density $\rho(x)$ and cross-sectional area A, and L is the pipe length. Note that for the axially FG pipe, $E(x)$ and $\rho(x)$ are functions of the axial coordinate x. In addition, we assume that the coefficient of thermal expansion and the thermal conductivity of the FG material are constant along the pipe, therefore, the influence of the thermal environment can be ignored. The following dimensionless variables are introduced

$$x^*=\frac{x}{L},\quad w^*=\frac{w}{L},\quad t^*=\frac{t}{L^2}\sqrt{\frac{E_0 I}{m_f+\rho_0 A}},\quad v^*=vL\sqrt{\frac{m_f}{E_0 I}}, \tag{11.2a–d}$$

$$\alpha(x^*)=\frac{E(x^*)}{E_0},\quad \beta=\sqrt{\frac{m_f}{m_f+\rho_0 A}},\quad \gamma(x^*)=\frac{m_f+\rho(x^*)A}{m_f+\rho_0 A}, \tag{11.2e–g}$$

where E_0 and ρ_0 are the corresponding Young's modulus and mass density at the end $x^*=0$. Substituting Eq. (11.2) into Eq. (11.1) gives the dimensionless equation (dropping the superposed asterisks for simplicity)

$$\frac{\partial^2}{\partial x^2}\left[\alpha(x)\frac{\partial^2 w}{\partial x^2}\right]+v^2\frac{\partial^2 w}{\partial x^2}+2\beta v\frac{\partial^2 w}{\partial x\partial t}+\gamma(x)\frac{\partial^2 w}{\partial t^2}=0, \tag{11.3a}$$

together with the boundary conditions

$$w(0,t)=0,\quad \frac{\partial w(0,t)}{\partial x}=0,\quad w(1,t)=0,\quad \frac{\partial w(1,t)}{\partial x}=0, \tag{11.3b-e}$$

The initial conditions are defined as follows:

$$w(x,0)=0,\quad \dot{w}(x,0)=v_0\sin(\pi x). \tag{11.4a,b}$$

11.3 Integral Transform Solution

To conduct the integral transform technique, the additional eigenvalue problem should be introduced for the governing equation (11.3a) with the boundary conditions (11.3b-e). The coordinate "x" is eliminated by using integral transformation, and the eigenvalue problem is employed for the transverse displacement as following:

$$\frac{\mathrm{d}^4 X_i(x)}{\mathrm{d}x^4} = \mu_i^4 X_i(x), \quad 0 < x < 1, \tag{11.5a}$$

with the boundary conditions

$$X_i(0) = 0, \quad \frac{\mathrm{d}X_i(0)}{\mathrm{d}x} = 0, \tag{11.5b,c}$$

$$X_i(1) = 0, \quad \frac{\mathrm{d}X_i(1)}{\mathrm{d}x} = 0, \tag{11.5d,e}$$

where $X_i(x)$ and μ_i are the eigenfunctions and eigenvalues of problem (11.5), respectively. The orthogonality property should be satisfied by the eigenfunctions:

$$\int_0^1 X_i(x) X_j(x) \mathrm{d}x = \delta_{ij} N_i, \tag{11.6}$$

with $\delta_{ij} = 0$ when $i \neq j$, and $\delta_{ij} = 1$ when $i = j$. The norm is defined as

$$N_i = \int_0^1 X_i^2(x) \mathrm{d}x. \tag{11.7}$$

Eigenvalue problem (11.5) can be solved analytically to generate

$$X_i(x) = \begin{cases} \dfrac{\cos[\mu_i(x - 1/2)]}{\cos(\mu_i/2)} - \dfrac{\cosh[\mu_i(x - 1/2)]}{\cosh(\mu_i/2)}, & \text{for } i \text{ odd}, \\ \dfrac{\sin[\mu_i(x - 1/2)]}{\sin(\mu_i/2)} - \dfrac{\sinh[\mu_i(x - 1/2)]}{\sinh(\mu_i/2)}, & \text{for } i \text{ even}, \end{cases} \tag{11.8a,b}$$

where the eigenvalues are calculated through the transcendental equations:

$$\tanh(\mu_i/2) = \begin{cases} -\tan(\mu_i/2), & \text{for } i \text{ odd}, \\ \tan(\mu_i/2), & \text{for } i \text{ even}, \end{cases} \tag{11.9a,b}$$

and the normalization of integral is

$$N_i = 1, \quad i = 1, 2, 3, \ldots \tag{11.10}$$

Then, the normalized eigenfunction yields

$$\tilde{X}_i(x) = \frac{X_i(x)}{N_i^{1/2}}. \tag{11.11}$$

For the transverse displacement, the integral transformation and the inversion equations are as follows:

$$\bar{w}_i(t) = \int_0^1 \tilde{X}_i(x) w(x,t) \mathrm{d}x, \quad \text{transform}, \tag{11.12a}$$

$$w(x,t) = \sum_{i=1}^{\infty} \tilde{X}_i(x) \bar{w}_i(t), \quad \text{inverse}. \tag{11.12b}$$

Now, the integral transformation procedure is conducted by operation of (11.3a) with $\int_0^1 \tilde{X}_i(x)\mathrm{d}x$, to obtain the transformed transverse displacement equation system:

$$\sum_{j=1}^{\infty} \left[A_{ij} \frac{\mathrm{d}^2 \bar{w}_j(t)}{\mathrm{d}t^2} + 2\beta v B_{ij} \frac{\mathrm{d}\bar{w}_j(t)}{\mathrm{d}t} + \left(v^2 C_{ij} + D_{ij} + 2E_{ij} + \mu_j^4 F_{ij}\right) \bar{w}_j(t) \right] = 0, \quad i = 1, 2, 3, \ldots, \tag{11.13a}$$

where the coefficients are given by the integrals below

$$A_{ij} = \int_0^1 \gamma(x) \tilde{X}_i(x) \tilde{X}_j(x) \mathrm{d}x, \tag{11.13b}$$

$$B_{ij} = \int_0^1 \tilde{X}_i(x) \tilde{X}_j(x)' \mathrm{d}x, \tag{11.13c}$$

$$C_{ij} = \int_0^1 \tilde{X}_i(x) \tilde{X}_j(x)'' \mathrm{d}x, \tag{11.13d}$$

$$D_{ij} = \int_0^1 \alpha(x)'' \tilde{X}_i(x) \tilde{X}_j(x)'' \mathrm{d}x, \tag{11.13e}$$

$$E_{ij} = \int_0^1 \alpha(x)' \tilde{X}_i(x) \tilde{X}_j(x)''' \mathrm{d}x, \tag{11.13f}$$

$$F_{ij} = \int_0^1 \alpha(x) \tilde{X}_i(x) \tilde{X}_j(x) \mathrm{d}x. \tag{11.13g}$$

In the similar way, the initial conditions can be also transformed to eliminate the "x" coordinate:

$$\bar{w}_i(0) = 0, \quad \frac{\mathrm{d}\bar{w}_i(0)}{\mathrm{d}t} = v_0 \int_0^1 \tilde{X}_i(x) \sin(\pi x) \mathrm{d}x, \quad i = 1, 2, 3, \ldots \tag{11.14a,b}$$

In the computational process, the expansion for the transverse displacement is truncated to finite orders NW. Equations (11.13) and (11.14) in the truncated series are calculated using the NDSolve in *Mathematica* (Wolfram 2003). Once the values of $\bar{w}_i$ are determined, the inversion formulas Eq. (11.12b) are subsequently employed to yield explicit expression for the transverse displacement $w(x, t)$.

11.4 Results and Discussion

11.4.1 Convergence Behavior of the Solution

We now present the convergence behavior of numerical results for the transverse displacement $w(x, t)$ of a functionally graded pipe conveying fluid calculated using the GITT approach. For the case examined, the geometrical parameters adopted by Zhai et al. (2011) are taken in Eq. (11.1): $L = 1010$ mm, $D = 22.85$ mm, and $d = 19.65$ mm, where D and d are the outer and inner diameters of pipe cross section, respectively. It is assumed that the material properties of the pipe, such as Young's modulus E and mass density ρ, vary continuously as a power law through the pipe axis (Alshorbagy et al. 2011): $P(x) = (P_L - P_R)(1 - x/L)^k + P_R$, where P_R and P_L are the corresponding material of the right and the left side of the pipe, and k is the non-negative power-law exponent which dictates the material variation profile through the pipe axis. In the following calculations, aluminum and zirconia are chosen for the corresponding material of the left and the right sides of the pipe, respectively, the material properties of which are (Huang and Li 2010): $E_a = 70$ GPa, $\rho_a = 2702$ kg/m^3 for aluminum and $E_z = 200$ GPa, $\rho_z = 5700$ kg/m^3 for zirconia. The fluid density conveying in the pipe is $\rho_f = 1000$ kg/m^3. The dimensionless variables can be obtained through Eq. (11.2). The solution of the system, Eqs. (11.13) and (11.14), is obtained with $NW \leq 16$ to analyze the convergence behavior.

Table 11.1 Convergence behavior of the dimensionless transverse displacement $w(x, t)$ of FG pipes conveying fluid for $v = 1.0$ and $k = 1.0$

x	$NW = 4$	$NW = 8$	$NW = 12$	$NW = 16$
$t = 10$				
0.1	0.0044244	0.0037878	0.0039159	0.0039563
0.3	0.0191635	0.0206171	0.0206690	0.0207072
0.5	0.0207420	0.0244299	0.0245583	0.0246031
0.7	0.0127285	0.0124189	0.0123877	0.0124042
0.9	0.0021788	0.0012293	0.0012868	0.0013476
$t = 20$				
0.1	−0.0060519	−0.0062440	−0.0060664	−0.0060803
0.3	−0.0291388	−0.0296250	−0.0298559	−0.0298584
0.5	−0.0358924	−0.0382805	−0.0382784	−0.0382978
0.7	−0.0235454	−0.0256259	−0.0257693	−0.0257332
0.9	−0.0040988	−0.0044978	−0.0042350	−0.0042428
$t = 30$				
0.1	0.0064297	0.0066167	0.0064907	0.0064724
0.3	0.0325025	0.0313819	0.0315394	0.0315556
0.5	0.0405411	0.0390336	0.0384158	0.0383750
0.7	0.0251618	0.0234384	0.0238947	0.0239246
0.9	0.0040959	0.0038549	0.0037553	0.0037439

The dimensionless transverse displacement $w(x, t)$ at different positions, $x = 0.1, 0.3, 0.5, 0.7$, and 0.9, of axially FG pipes conveying fluid is presented in Tables 11.1 and 11.2, respectively. The convergence behavior of the integral transform solution is examined for increasing truncation terms $NW = 4, 8, 12$ and 16 at $t = 10, 20$, and 30, respectively. For the dimensionless transverse displacement with $v = 1.0$ and $k = 1.0$, it can be observed that convergence is achieved essentially with a reasonably low truncation order ($N \leq 8$). For a full convergence to three significant digits, more terms (e.g., $N \leq 12$) are required. The results at $t = 30$ indicate that the excellent convergence behavior of the integral transform solution does not change with time, verifying the good long-time numerical stability of the scheme. For the dimensionless transverse displacement with $v = 3.0$ and $k = 1.0$, convergence to three significant digits is achieved with truncation order $N \leq 16$, which demonstrates that the increasing of v can make the solution with a relatively slow convergence. For the same cases, the profiles of the transverse displacement at $t = 20$ are illustrated in Fig. 11.2 with different truncation orders, where it can be clearly seen that the convergence behavior of the integral transform solution for the case of $v = 1.0$ is better than the case of $v = 3.0$.

Table 11.2 Convergence behavior of the dimensionless transverse displacement $w(x, t)$ of FG pipes conveying fluid for $v = 3.0$ and $k = 1.0$

x	$NW = 4$	$NW = 8$	$NW = 12$	$NW = 16$
$t = 10$				
0.1	0.0030463	0.0024006	0.0026071	0.0026145
0.3	0.0083268	0.0100507	0.0099890	0.0099936
0.5	0.0025960	0.0057039	0.0059138	0.0059497
0.7	−0.0013099	−0.0021990	−0.0022931	−0.0023027
0.9	−0.0006775	−0.0011992	−0.0010060	−0.0010314
$t = 20$				
0.1	−0.0019432	−0.0013335	−0.0015234	−0.0015166
0.3	−0.0071319	−0.0088004	−0.0088014	−0.0088167
0.5	−0.0055658	−0.0118234	−0.0123907	−0.0124569
0.7	−0.0028723	−0.0073600	−0.0070684	−0.0071240
0.9	−0.0005661	−0.0009220	−0.0011155	−0.0011267
$t = 30$				
0.1	0.0033760	0.0043476	0.0041930	0.0041604
0.3	0.0113849	0.0171511	0.0168645	0.0169404
0.5	0.0093804	0.0150589	0.0157554	0.0157545
0.7	0.0054698	0.0066738	0.0066753	0.0066903
0.9	0.0008500	0.0005556	0.0008388	0.0009427

11.4.2 Verification of the Solution

The influence of fluid velocity on the first five dimensionless natural frequencies of axially FG pipes conveying fluid with the power-law exponent $k = 1.0$ is presented in Table 11.3. To obtain the natural circular frequencies for the transverse vibration of the system, the coupled ODEs, Eq. (11.13), can be represented in the matrix form as follows:

$$\mathbf{M}\ddot{\mathbf{w}}(t) + \mathbf{C}\dot{\mathbf{w}}(t) + \mathbf{K}\mathbf{w}(t) = \mathbf{F}(t). \tag{11.15}$$

The fluid velocities of 0, 10, 20, 30, 40, and 50 m/s are considered, and with the increasing of the velocity, all of the five natural frequencies of the system decrease. To demonstrate the validity and accuracy of the proposed GITT approach, GITT solution for dimensionless free vibration frequencies for clamped–clamped pipes with $m_f = 0$ and $k = 0$ is calculated to compare with the results presented in the literature (Leissa and Qatu 2011), where the excellent agreement between them can be found, as shown in Table 11.3.

Fig. 11.2 GITT solutions with different truncation orders NW for the dimensionless transverse displacement profiles (**a**) $w(x,t)|_{t=20}$ for $v = 1.0$ and (**b**) $w(x,t)|_{t=20}$ for $v = 3.0$ of FG pipes conveying fluid

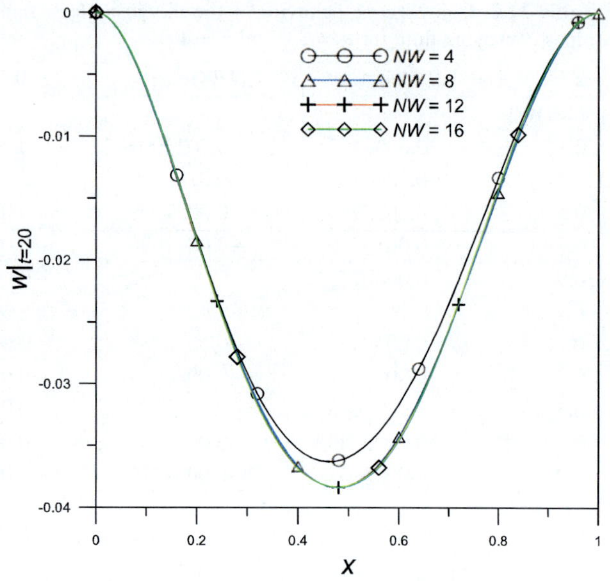

(a)

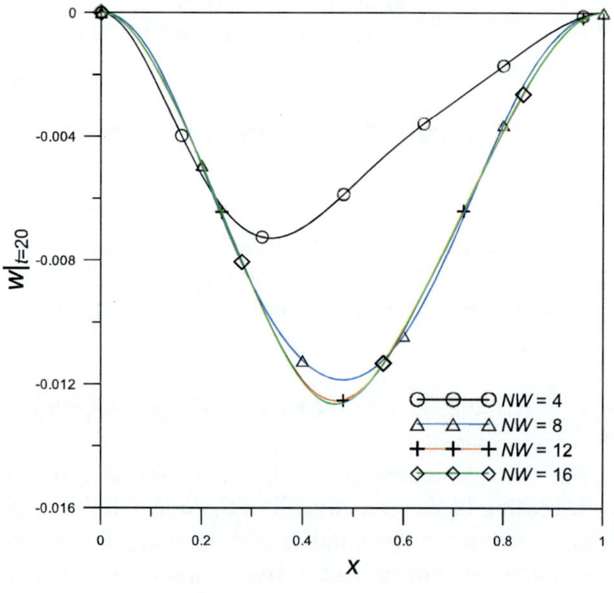

(b)

Table 11.3 Influence of fluid velocity on the first five dimensionless natural frequencies of FG pipes conveying fluid ($k = 1.0$)

fluid velocity (m/s)	ω_1	ω_2	ω_3	ω_4	ω_5
0	22.373[a]	61.673[a]	120.903[a]	199.859[a]	298.556[a]
0	22.373[b]	61.673[b]	120.903[b]	199.859[b]	298.556[b]
0	26.866	74.323	145.936	241.576	361.572
10	26.850	74.306	145.918	241.558	361.556
20	26.801	74.255	145.865	241.506	361.508
30	26.720	74.171	145.777	241.419	361.428
40	26.605	74.052	145.653	241.296	361.315
50	26.457	73.899	145.493	241.139	361.171

[a]Dimensionless free vibration frequencies for clamped–clamped beams (Leissa and Qatu 2011)
[b]GITT solution for dimensionless free vibration frequencies for clamped–clamped pipes ($m_f = 0$ and $k = 0$)

11.4.3 Parametric Study

In this section, transverse displacement of axially FG pipes conveying fluid with clamped–clamped boundary conditions is analyzed to illustrate the applicability of the proposed approach. Different values of the mass ratio β, the Young's modulus ratio E_{ratio}, the material distribution k, the dimensionless flow velocity v are chosen to assess their effects on the dynamic behavior of the system. In the following analysis, we use a relative high truncation order, $NW = 16$, for a sufficient accuracy.

11.4.3.1 The Effect of Young's Modulus Variation

The first three dimensionless natural frequencies of the axially FG pipe conveying fluid for different Young's modulus ratios, power-law exponent, and mass ratios are tabulated in Tables 11.4, 11.5, and 11.6, where the following parameters are adopted: $E_l = 70\,\text{GPa}$, $E_{ratio} = E_L/E_R$, $\rho_f = 1000\,\text{kg/m}^3$, $\rho_{ratio} = \rho_L/\rho_R = 1.0$ and $\rho_L/\rho_f = 4.0$. The mass ratio is calculated by Eq. (11.2) with the specified value of pipe thickness t and $d = 19.65\,\text{mm}$. The flow velocity of the fluid is 20 m/s. For the specified modulus ratio and power exponent, all of the first three dimensionless natural frequencies increase with the decrease of the mass ratio. The effect of the modulus ratios on the fundamental frequency of fluid-conveying pipe with the mass ratio of 0.796 is presented in Fig. 11.3. It is observed that, the fundamental frequency decreases significantly with increasing of modulus ratio especially for large power exponent. On the other hand, no significant changes can be seen in the fundamental frequency for different modulus ratios for lower value of power exponent. Figure 11.4 illustrates the variation of the fundamental frequency with the power exponent for fluid-conveying pipe with the mass ratio of 0.796, which shows that the increase in power exponent causes the increase in frequency for $E_{ratio} < 1$, the decrease in frequency for $E_{ratio} > 1$ and no changes occur for $E_{ratio} = 1$.

Table 11.4 The variation of the first natural angular frequencies for different material distributions k and mass ratios β when flow velocity is 20 m/s, $E_{ratio} = E_l/E_r$, $\rho_{ratio} = \rho_l/\rho_r = 1.0$ and $\rho_l/\rho_f = 4.0$

t (mm)	β	E_{ratio}	$k = 0.0$	$k = 0.1$	$k = 0.2$	$k = 0.5$	$k = 1.0$	$k = 2.0$	$k = 5.0$	$k = 10.0$
1.0	0.796	0.25	22.1909	25.2702	27.1859	30.5936	33.4550	36.1690	39.0072	40.4687
		0.50	22.1909	23.3497	24.1498	25.6193	26.8588	28.0211	29.1967	29.7777
		1.00	22.1909	22.1909	22.1909	22.1909	22.1909	22.1909	22.1909	22.1909
		2.00	22.1909	21.5289	20.9829	19.8618	18.8849	17.9785	17.0936	16.6791
		4.00	22.1909	21.1690	20.2790	18.3220	16.5444	14.9059	13.3283	12.5955
2.0	0.672	0.25	22.3008	25.3695	27.2788	30.6755	33.5282	36.2347	39.0667	40.5260
		0.50	22.3008	23.4554	24.2525	25.7159	26.9498	28.1068	29.2778	29.8571
		1.00	22.3008	22.3008	22.3008	22.3008	22.3008	22.3008	22.3008	22.3008
		2.00	22.3008	21.6413	21.0977	19.9830	19.0138	18.1167	17.2418	16.8313
		4.00	22.3008	21.2828	20.3968	18.4519	16.6913	15.0763	13.5273	12.8067
3.0	0.587	0.25	22.3333	25.3989	27.3063	30.6998	33.5499	36.2543	39.0844	40.5430
		0.50	22.3333	23.4867	24.2829	25.7445	26.9767	28.1322	29.3018	29.8807
		1.00	22.3333	22.3333	22.3333	22.3333	22.3333	22.3333	22.3333	22.3333
		2.00	22.3333	21.6746	21.1317	20.0189	19.0520	18.1575	17.2856	16.8762
		4.00	22.3333	21.3165	20.4318	18.4904	16.7348	15.1266	13.5858	12.8688

Table 11.5 The variation of the second natural angular frequencies for different material distributions k and mass ratios β when flow velocity is 20 m/s, $E_{ratio} = E_l/E_r$, $\rho_{ratio} = \rho_l/\rho_r = 1.0$ and $\rho_l/\rho_f = 4.0$

t (mm)	β	E_{ratio}	$k = 0.0$	$k = 0.1$	$k = 0.2$	$k = 0.5$	$k = 1.0$	$k = 2.0$	$k = 5.0$	$k = 10.0$
1.0	0.796	0.25	61.5004	69.4713	74.6411	84.2515	92.8974	101.469	110.495	114.810
		0.50	61.5004	64.4750	66.6003	70.6875	74.4028	78.0533	81.7840	83.5138
		1.00	61.5004	61.5004	61.5004	61.5004	61.5004	61.5004	61.5004	61.5004
		2.00	61.5004	59.8213	58.4126	55.3901	52.5094	49.6838	46.9156	45.7453
		4.00	61.5004	58.9166	56.6441	51.4604	46.2746	41.1824	36.2901	34.2703
2.0	0.672	0.25	61.5959	69.5577	74.7221	84.3233	92.9617	101.527	110.547	114.860
		0.50	61.5959	64.5668	66.6895	70.7715	74.4821	78.1281	81.8547	83.5830
		1.00	61.5959	61.5959	61.5959	61.5959	61.5959	61.5959	61.5959	61.5959
		2.00	61.5959	59.9190	58.5124	55.4955	52.6216	49.8040	47.0441	45.8770
		4.00	61.5959	59.0156	56.7467	51.5737	46.4032	41.3315	36.4620	34.4515
3.0	0.587	0.25	61.6276	69.5864	74.7490	84.3471	92.9830	101.546	110.564	114.876
		0.50	61.6276	64.5973	66.7191	70.7995	74.5085	78.1529	81.8782	83.6060
		1.00	61.6276	61.6276	61.6276	61.6276	61.6276	61.6276	61.6276	61.6276
		2.00	61.6276	59.9515	58.5456	55.5305	52.6589	49.8439	47.0868	45.9207
		4.00	61.6276	59.0485	56.7809	51.6114	46.4459	41.3809	36.5192	34.5117

Table 11.6 The variation of the third natural angular frequencies for different material distributions k and mass ratios β when flow velocity is 20 m/s, $E_{ratio} = E_l/E_r$, $\rho_{ratio} = \rho_l/\rho_r = 1.0$ and $\rho_l/\rho_f = 4.0$

t (mm)	β	E_{ratio}	$k = 0.0$	$k = 0.1$	$k = 0.2$	$k = 0.5$	$k = 1.0$	$k = 2.0$	$k = 5.0$	$k = 10.0$
1.0	0.796	0.25	120.726	135.934	145.948	164.997	182.693	200.618	219.333	227.933
		0.50	120.726	126.374	130.466	138.522	146.073	153.635	161.360	164.876
		1.00	120.726	120.726	120.726	120.726	120.726	120.726	120.726	120.726
		2.00	120.726	117.564	114.895	109.017	103.185	97.3618	91.6997	89.3743
		4.00	120.726	115.872	111.598	101.632	91.1682	80.6355	70.6481	66.6398
2.0	0.672	0.25	120.822	136.020	146.028	165.068	182.757	200.677	219.386	227.984
		0.50	120.822	126.466	130.555	138.606	146.153	153.711	161.432	164.946
		1.00	120.822	120.822	120.822	120.822	120.822	120.822	120.822	120.822
		2.00	120.822	117.663	114.996	109.124	103.298	97.4820	91.8275	89.5047
		4.00	120.822	115.972	111.702	101.747	91.2969	80.7826	70.8185	66.8177
3.0	0.587	0.25	120.855	136.049	146.056	165.092	182.779	200.697	219.405	228.002
		0.50	120.855	126.498	130.586	138.635	146.180	153.737	161.457	164.970
		1.00	120.855	120.855	120.855	120.855	120.855	120.855	120.855	120.855
		2.00	120.855	117.697	115.031	109.160	103.336	97.5231	91.8713	89.5494
		4.00	120.855	116.006	111.738	101.786	91.3409	80.8330	70.8770	66.8789

Fig. 11.3 Variation of the dimensionless frequency (ω_1) with Young's modulus ratio

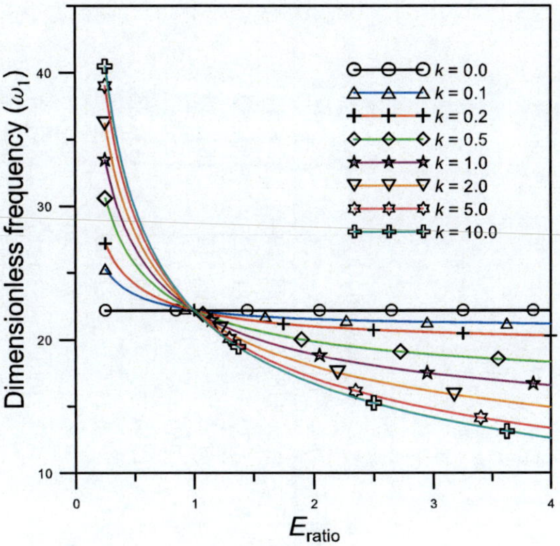

Fig. 11.4 Variation of the dimensionless frequency (ω_1) with power exponent (k)

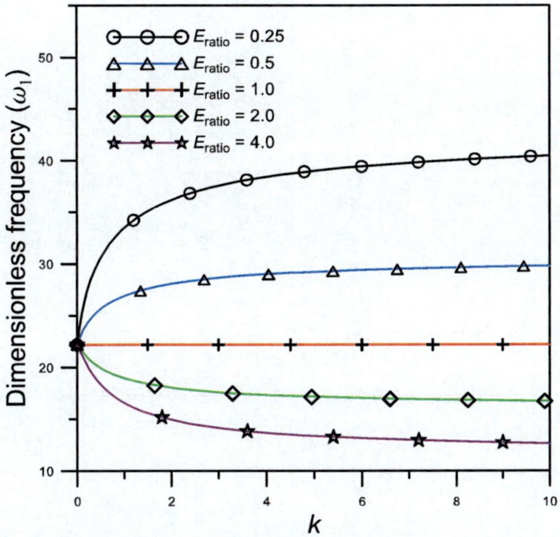

11.4.3.2 The Effect of Material Distribution

To examine the effect of material distribution on the frequencies of the axially FG pipe conveying fluid, the integral transform solutions are obtained based on the material properties given in Sect. 11.4.1 and the fluid velocity of 20 m/s. The first five natural angular frequencies for different mass ratios and power exponents are reported in Table 11.7. All of the natural frequencies ($\omega_1, \omega_2, \ldots, \omega_5$) increase with the power exponent for the specified mass ratio. Note that the natural frequencies increase with the decrease of the mass ratio for the case of $k = 0.0$, which means the pipe is made of single-component, aluminum. However, when considering the

Table 11.7 The variation of natural angular frequencies for different material distributions k and mass ratios β when flow velocity is 20 m/s

t (mm)	β	ω_i	$k = 0.0$	$k = 0.1$	$k = 0.2$	$k = 0.5$	$k = 1.0$	$k = 2.0$	$k = 5.0$	$k = 10.0$
1.0	0.796	$i = 1$	22.1835	23.8868	24.9026	26.4715	27.4513	28.1186	28.8294	29.4111
		$i = 2$	61.5036	65.7479	68.3906	72.7928	76.1318	79.0050	81.8223	83.2525
		$i = 3$	120.732	128.668	133.700	142.428	149.573	156.161	162.514	165.285
		$i = 4$	199.693	212.486	220.721	235.305	247.679	259.404	270.748	275.362
		$i = 5$	298.392	317.258	329.420	351.361	370.905	389.776	407.194	413.686
2.0	0.672	$i = 1$	22.2976	23.8315	24.6962	25.9020	26.4864	26.7497	27.1061	27.5738
		$i = 2$	61.5972	65.3592	67.5855	71.0306	73.3695	75.2171	77.0180	78.0331
		$i = 3$	120.825	127.785	131.993	138.872	144.125	148.770	153.121	155.007
		$i = 4$	199.781	210.932	217.798	229.340	238.617	247.159	255.195	258.328
		$i = 5$	298.477	314.839	324.951	342.322	357.127	371.143	383.736	388.144
3.0	0.587	$i = 1$	22.3317	23.7692	24.5496	25.5605	25.9443	26.0152	26.2112	26.6270
		$i = 2$	61.6283	65.1117	67.1047	70.0359	71.8650	73.2123	74.5377	75.3643
		$i = 3$	120.856	127.252	131.000	136.890	141.179	144.870	148.272	149.752
		$i = 4$	199.812	210.013	216.115	226.033	233.735	240.712	247.171	249.620
		$i = 5$	298.508	313.428	322.400	337.340	349.740	361.370	371.651	375.089

Table 11.8 The variation of vibration amplitudes for different material distributions k and mass ratios β when flow velocity is 20 m/s

t (mm)	β	$k = 0.0$	$k = 0.1$	$k = 0.2$	$k = 0.5$	$k = 1.0$	$k = 2.0$	$k = 5.0$	$k = 10.0$
1.0	0.796	0.05064	0.04742	0.04562	0.04265	0.04067	0.03966	0.03864	0.03787
2.0	0.672	0.05071	0.04779	0.04607	0.04365	0.04226	0.04176	0.04093	0.04042
3.0	0.587	0.05076	0.04794	0.04657	0.04431	0.04315	0.04276	0.04237	0.04192

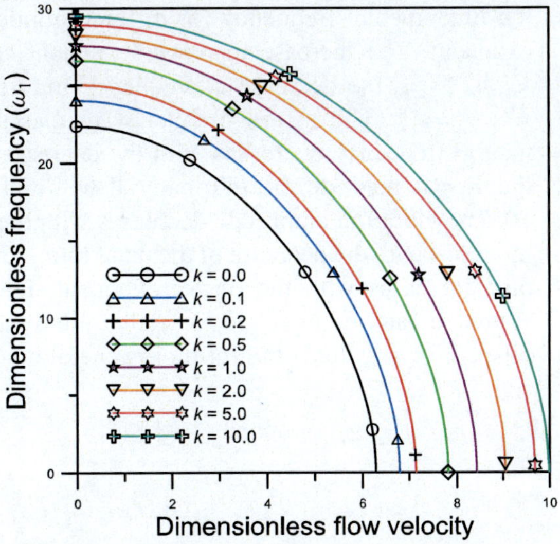

Fig. 11.5 Variation of the dimensionless frequency (ω_1) with dimensionless flow velocity (v)

FG material with the power exponents $k = 0.1, 0.2, \ldots, 10$, the natural frequencies decrease with the decrease of the mass ratio.

The variation of vibration amplitudes for different material distributions and mass ratios is listed in Table 11.8. It can be seen that the vibration amplitude decreases with the increase of the power exponent and increases with the decrease of the mass ratio.

In addition, the effect of material distribution on the critical velocity of fluid-conveying pipe with the material properties given in Sect. 11.4.1 is analyzed, as shown in Fig. 11.5, which exhibits the variation of the fundamental frequency with the flow velocity of fluid for different power exponents. It can be clear seen that the fundamental frequency decreases with the flow velocity for the specified power exponent, and the critical velocity ($\omega_1 = 0$) increases with the power exponent.

11.5 Conclusions

The generalized integral transform technique (GITT) has proved in this chapter to be a good approach for the analysis of dynamic behavior of an axially FG pipe conveying fluid, providing an accurate numerical–analytical solution for the natural frequencies and transverse displacements. The investigation shows that the solutions converge to the values with three significant figures at a reasonable low truncation order $N \leq 12$ for $v = 1.0$, and the increasing of v can make the solution with a relatively slow convergence. The numerical results obtained are in good agreement with the ones presented in the literature. The parametric studies indicate that the fundamental frequency decreases significantly with increasing of modulus ratio especially for large power exponent, while no significant changes can be seen in the fundamental frequency for different modulus ratios for lower value of power exponent. The increase in power exponent causes the increase in frequency for $E_{ratio} < 1$, the decrease in frequency for $E_{ratio} > 1$ and no changes occur for $E_{ratio} = 1$. The natural frequency increases with the power exponent, and the natural frequency decreases with the decrease of the mass ratio when considering the pipe is made of the FG material with the power exponents $k = 0.1, 0.2, \ldots, 10$. The vibration amplitude decreases with the increase of the power exponent and increases with the decrease of the mass ratio. The critical velocity of fluid-conveying pipe increases with the power exponent. For future investigation, the proposed approach can be employed to predict the dynamic behavior of a transversally FG pipe conveying fluid, and for more general boundary conditions.

Part III
Thermal Analysis of Multilayer Pipelines

Keeping the thermodynamic state of produced fluid within adequate range is essential to meet the production requirements and minimize downtime due to possible pipeline blockage. For a given combination of hydrocarbons flowing into a multilayer pipeline, the range of pressure variation is largely determined by the wellhead and separator pressures, while the temperature distribution of the produced fluid is determined by the energy balance over the pipeline. For the thermal design of steady-state operation conditions, adequate thermal insulation system must be specified to meet the requirement of keeping the temperature above the wax appearance temperature. Under shut-in conditions, a reasonable long cooling-down time should be achieved to prevent the formation of gas hydrate.

Part III of the book deals with the thermal analysis of multilayer composite pipelines. In Chap. 12, basic governing equations for thermal analysis of pipelines are presented, together with the overall heat transfer coefficient for multilayered composite pipeline and the temperature distribution of the produced fluid along the pipeline under steady-state conditions. Chapter 13 presents a global thermal analysis of multilayer composite pipelines and establishes the requirement of active heating when the passive thermal insulation cannot meet the thermal design requirement. Active heating by hot medium circulation and direct electrical heating are discussed. The challenge of heavy oil production is addressed in Chap. 14. Chapter 15 analyzes direct electrical heating at steady-state operation in more details. Finally, Chap. 16 presents transient thermal analysis of multilayer composite pipelines with direct electrical heating, using finite difference method and lumped models, respectively.

Chapter 12
Fundamentals of Thermal Analysis

The objective of thermal design of deepwater pipelines is to keep the produced hydrocarbons in a designed thermodynamic state to avoid the occurrence of undesired thermodynamic or transport processes, such as gas hydrate formation and paraffin deposition in the pipeline.

12.1 Equation of Energy

The foundation of thermal analysis of pipelines is the first law of thermodynamics, that is, the law of conservation of energy, which can be expressed in the following rate form:

$$\frac{\mathrm{d}E}{\mathrm{d}t} = \dot{Q} - \dot{W}, \tag{12.1}$$

where E is the total energy of a system, $\dot{Q}$ is the rate of heat transfer from the environment into the system through the boundary surface of the system, and $\dot{W}$ is the rate of the work done by the system to the environment.

We first separate the total energy E of system into internal energy U and energy of other form E_g,

$$\frac{\mathrm{d}(U + E_g)}{\mathrm{d}t} = \dot{Q} - \dot{W}, \tag{12.2}$$

or

$$\frac{\mathrm{d}U}{\mathrm{d}t} = \dot{Q} - \dot{W} - \frac{\mathrm{d}E_g}{\mathrm{d}t}. \tag{12.3}$$

C. An et al., *Structural and Thermal Analyses of Deepwater Pipes*,
https://doi.org/10.1007/978-3-030-53540-7_12

The rate of change of the other forms of energy is expressed as a heat generation term:

$$\dot{Q}_g = -\frac{\mathrm{d}E_g}{\mathrm{d}t}. \tag{12.4}$$

Thus, the first law of thermodynamics for a system can be written as

$$\frac{\mathrm{d}U}{\mathrm{d}t} = \dot{Q} + \dot{Q}_g - \dot{W}. \tag{12.5}$$

For flowing system with one inlet and one outlet such as a pipeline, we have the following equation for mass conservation:

$$\frac{\mathrm{d}M}{\mathrm{d}t} = \dot{m}_{in} - \dot{m}_{out}, \tag{12.6}$$

where M is the total mass in the open system or control volume, $\dot{m}_{in}$ is the mass flow rate into the control volume, and $\dot{m}_{out}$ is the mass flow rate exiting from the control volume.

For steady-state flow system, we have

$$\frac{\mathrm{d}M}{\mathrm{d}t} = 0. \tag{12.7}$$

That is

$$\dot{m}_{in} = \dot{m}_{out} = \dot{m}. \tag{12.8}$$

For a flow system, the first law of thermodynamics is written as

$$\frac{\mathrm{d}U}{\mathrm{d}t} = \dot{Q} + \dot{Q}_g + \dot{m}_{in}u_{in} - \dot{m}_{out}u_{out} + p_{in}A_{in}V_{in} - p_{out}A_{out}V_{out}, \tag{12.9}$$

where u is the specific internal energy of the fluid, p is the pressure, A is cross-sectional area, and V is the velocity of the fluid. The other forms of work are neglected. Using

$$\dot{m} = \rho A V, \tag{12.10}$$

and

$$h = u + \frac{p}{\rho}, \tag{12.11}$$

where h is the specific enthalpy of the fluid, we rewrite the first law of the flowing system as:

$$\frac{\mathrm{d}U}{\mathrm{d}t} = \dot{Q} + \dot{Q}_g + \dot{m}_{in}h_{in} - \dot{m}_{out}h_{out}. \tag{12.12}$$

12.2 Heat Conduction in Pipe Wall

12.2.1 Heat Conduction in a Single-Layer Pipe Wall

In this subsection, we consider one-dimensional, steady-state heat conduction in a single-layer pipe with convective boundary conditions at both inner and outer pipe walls, as illustrated in Fig. 12.1. The thermal conductivity of the pipe material k is considered constant. There is no volumetrical heat generation in the pipe.

The governing heat conduction equation is written in cylindrical coordinate system as:

$$\frac{k}{r}\frac{\mathrm{d}}{\mathrm{d}r}\left(r\frac{\mathrm{d}T}{\mathrm{d}r}\right) = 0, \quad r_1 < r < r_o, \tag{12.13a}$$

with boundary conditions:

$$k\frac{\mathrm{d}T}{\mathrm{d}r} = h_1\,(T - T_f), \quad \text{at} \quad r = r_1, \tag{12.13b}$$

$$-k\frac{\mathrm{d}T}{\mathrm{d}r} = h_o\,(T - T_a), \quad \text{at} \quad r = r_o, \tag{12.13c}$$

where h_1 is the convective heat transfer coefficient at inner pipe wall r_1, h_o is the convective heat transfer coefficient at outer pipe wall r_o, and T_f and T_a are, respectively, the inner and outer fluid temperatures.

Equation (12.13a) is integrated in r to get the heat flux:

$$q_r'' = -k\frac{\mathrm{d}T}{\mathrm{d}r} = \frac{A_1}{r}. \tag{12.14}$$

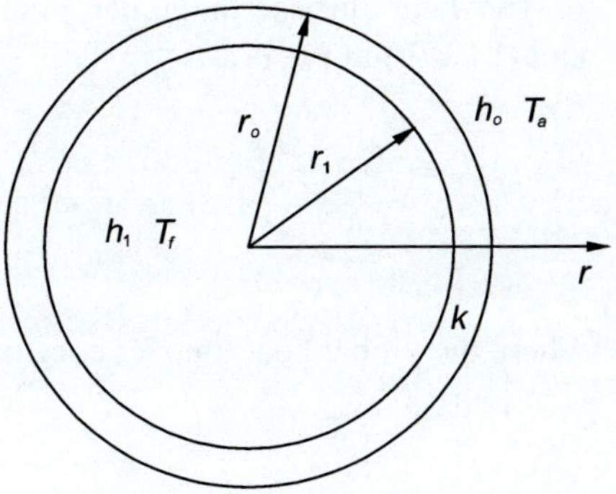

Fig. 12.1 Cross section of a single-layer pipe

Equation (12.14) is integrated again in r to get the temperature distribution:

$$T(r) = -\frac{A_1}{k} \ln r + A_2 + T_a. \tag{12.15}$$

It should be noted that for future convenience we separate the outside fluid temperature T_a from the second integration constant A_2.

Equations (12.14) and (12.15) are used in the boundary conditions, Eqs. (12.13b) and (12.13c):

$$-\frac{A_1}{r_1} = h_1 (-\frac{A_1}{k} \ln r_1 + A_2 + T_a - T_f), \tag{12.16a}$$

$$\frac{A_1}{r_o} = h_o (-\frac{A_1}{k} \ln r_o + A_2). \tag{12.16b}$$

Equations (12.16a) and (12.16b) are solved for A_1 and A_2:

$$A_1 = \frac{T_f - T_a}{\frac{1}{r_1 h_1} + \frac{1}{k} \ln \frac{r_o}{r_1} + \frac{1}{r_o h_o}}, \tag{12.17a}$$

$$A_2 = \left(\frac{\ln r_o}{k} + \frac{1}{r_o h_o}\right) \frac{T_f - T_a}{\frac{1}{r_1 h_1} + \frac{1}{k} \ln \frac{r_o}{r_1} + \frac{1}{r_o h_o}}. \tag{12.17b}$$

The temperature distribution in the single-layer pipe is thus obtained as

$$T(r) = -\frac{T_f - T_a}{\frac{1}{r_1 h_1} + \frac{1}{k} \ln \frac{r_o}{r_1} + \frac{1}{r_o h_o}} \frac{\ln r}{k} + \frac{T_f - T_a}{\frac{1}{r_1 h_1} + \frac{1}{k} \ln \frac{r_o}{r_1} + \frac{1}{r_o h_o}} \left(\frac{\ln r_o}{k} + \frac{1}{r_o h_o}\right) + T_a \tag{12.18}$$

$$= \frac{T_f - T_a}{\frac{1}{r_1 h_1} + \frac{1}{k} \ln \frac{r_o}{r_1} + \frac{1}{r_o h_o}} \left(\frac{1}{k} \ln \frac{r_o}{r} + \frac{1}{r_o h_o}\right) + T_a.$$

The heat flux at the inner pipe wall r_1 is obtained by using Eqs. (12.17a) and (12.17b) in Eq. (12.19):

$$q''_{r_1} = \frac{T_f - T_a}{\frac{1}{r_1 h_1} + \frac{1}{k} \ln \frac{r_o}{r_1} + \frac{1}{r_o h_o}} \frac{1}{r_1} = \frac{T_f - T_a}{\frac{1}{h_1} + \frac{r_1}{k} \ln \frac{r_o}{r_1} + \frac{r_1}{r_o h_o}} = U\,(T_f - Ta), \tag{12.19}$$

where the global heat transfer coefficient U is given by

$$U = \frac{1}{\frac{1}{h_1} + \frac{r_1}{k} \ln \frac{r_o}{r_1} + \frac{r_1}{r_o h_o}}. \tag{12.20}$$

Fig. 12.2 Cross section of a M layer composite pipeline

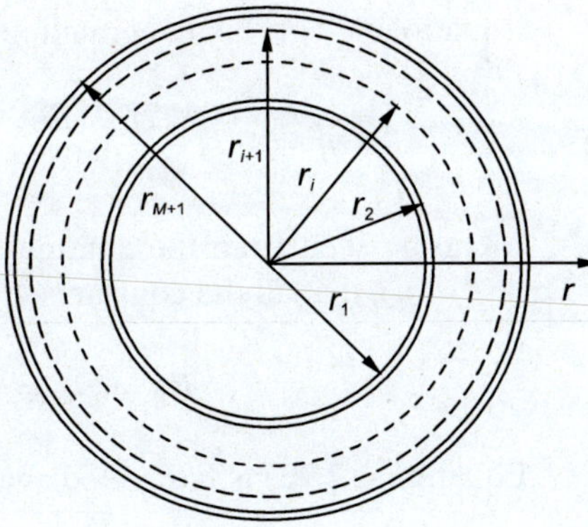

12.2.2 Heat Conduction in a Multilayer Pipe Wall

In this subsection, we extend the one-dimensional, steady-state heat conduction in a single-layer pipe to a multilayer composite pipe, consisted of M layers, as illustrated in Fig. 12.2. The inner and outer radii of i-th layer are r_i and r_{i+1}, respectively. The thermal conductivity of the i-th layer material k_i is considered constant. There is no heat generation in any of the layers. The pipe is subject to convective heat transfer at the inner pipe wall r_1 and the outer pipe wall $r_o = r_{M+1}$, with h_1 being the convective heat transfer coefficient at inner pipe wall, h_o the convective heat transfer coefficient at outer wall, and T_f and T_a, respectively, the inner and outer fluid temperatures.

The governing heat conduction equation is written as

$$\frac{k_i}{r}\frac{\mathrm{d}}{\mathrm{d}r}\left(r\frac{\mathrm{d}T_i}{\mathrm{d}r}\right) = 0, \quad r_i < r < r_{i+1}, \quad \text{for} \quad i = 1, \ldots, M \tag{12.21a}$$

with inner and outer boundary conditions:

$$k_1\frac{\mathrm{d}T_1}{\mathrm{d}r} = h_1\,(T_1 - T_f), \quad \text{at} \quad r = r_1, \tag{12.21b}$$

$$-k_M\frac{\mathrm{d}T_M}{\mathrm{d}r} = h_o\,(T_M - T_a), \quad \text{at} \quad r = r_o, \tag{12.21c}$$

and interface boundary conditions:

$$k_i\frac{\mathrm{d}T_i}{\mathrm{d}r} = k_{i+1}\frac{\mathrm{d}T_{i+1}}{\mathrm{d}r}, \quad \text{at} \quad r = r_{i+1}, \quad \text{for} \quad i = 1, \ldots, M-1, \tag{12.21d}$$

$$T_i = T_{i+1}, \quad \text{at} \quad r = r_{i+1}, \quad \text{for} \quad i = 1, \ldots, M-1, \tag{12.21e}$$

where T_i is the temperature in i-th layer of the pipe.

Equation (12.21a) is integrated in r to get the heat flux:

$$q_i'' = -k_i \frac{\mathrm{d}T_i}{\mathrm{d}r} = \frac{B_i}{r}, \quad \text{for} \quad i = 1, \ldots, M. \tag{12.22}$$

It can be seen from the continuity condition of the heat flux at the interfaces, Eq. (12.21d), that all the constants B_i take the same value:

$$B_1 = B_2 = \ldots = B_i = \ldots = B_M. \tag{12.23}$$

Equation (12.22) is integrated again in r to get the temperature distribution:

$$T_i(r) = -\frac{B_1}{k_i} \ln r + C_i + T_a, \quad \text{for} \quad i = 1, \ldots, M. \tag{12.24}$$

Equations (12.22), (12.23), and (12.24) are used in boundary conditions, Eqs. (12.21b) and (12.21c), and the temperature continuity conditions at the interfaces, Eqs. (12.21e)

$$-\frac{B_1}{r_1} = h_1 \left(-\frac{B_1}{k_1} \ln r_1 + C_1 + T_a - T_f\right), \tag{12.25}$$

$$\frac{B_1}{r_o} = h_o \left(-\frac{B_1}{k_M} \ln r_o + C_m\right). \tag{12.26}$$

$$-\frac{B_1}{k_i} \ln r_{i+1} + C_i + T_a = -\frac{B_1}{k_{i+1}} \ln r_{i+1} + C_{i+1} + T_a,$$
$$\text{at } r = r_{i+1}, \quad \text{for} \quad i = 1, \ldots, M-1. \tag{12.27}$$

The system of $M + 1$ linear equations formed by Eqs. (12.25), (12.26), and (12.27) is solved for the $M + 1$ unknown constants, B_1 and C_i, $i = 1, \ldots, M$:

$$B_1 = \frac{T_f - T_a}{\frac{1}{r_1 h_1} + \sum_{i=1}^{M} \frac{1}{k_i} \ln \frac{r_{i+1}}{r_i} + \frac{1}{r_o h_o}}, \tag{12.28a}$$

$$C_M = \left(\frac{\ln r_o}{k} + \frac{1}{r_o h_o}\right) B_1 \tag{12.28b}$$

$$C_i = C_{i+1} + B_1 \ln r_{i+1} \left(\frac{1}{k_i} - \frac{1}{k_{i+1}}\right), i = M-1, \ldots, 1. \tag{12.28c}$$

The heat flux at the inner pipe wall r_1 is obtained by using Eqs. (12.17a) and (12.17b) in Eq. (12.22):

$$q_{r_1}'' = \frac{B_1}{r_1} = \frac{T_f - T_a}{\frac{1}{r_1 h_1} + \sum_{i=1}^{M} \frac{1}{k_i} \ln \frac{r_{i+1}}{r_i} + \frac{1}{r_o h_o}} \frac{1}{r_1} = U\,(T_f - Ta), \tag{12.29}$$

where the global heat transfer coefficient U for a M-layer composite pipe is given by

$$U = \frac{1}{\frac{1}{h_1} + \sum_{i=1}^{M} \frac{r_1}{k_i} \ln \frac{r_{i+1}}{r_i} + \frac{r_1}{r_o h_o}}. \tag{12.30}$$

12.3 Steady-State Temperature Distribution of Produced Fluid in an Unheated Pipeline

Let us consider the temperature distribution of produced fluid in an unheated pipeline. At steady state, the one-dimensional energy transport in the produced fluid is uncoupled from the heat conduction in the pipeline wall, and governed by the following equation:

$$\dot{m}\, c_{pf} \frac{\mathrm{d}T_f}{\mathrm{d}s} = -2\pi r_1 U (T_f - Ta), \quad \text{for} \quad 0 < s < L, \tag{12.31}$$

with the inlet condition,

$$T_f(0) = T_{f,in}, \quad \text{at} \quad s = 0, \tag{12.32}$$

where $\dot{m}$ is the mass flow rate of the produced fluid, c_{pf} the specific heat, and $T_{f,in}$ the temperature of the produced fluid at the inlet of the pipeline. The global heat transfer coefficient U for a M-layer composite pipe is given by Eq. (12.30).

Equation (12.31) is readily solved with Eq. (12.32) to obtain the temperature distribution of the produced fluid along the pipeline:

$$T_f(s) = T_a + (T_{f,in} - T_a) \exp\left(-\frac{2\pi r_1 U\, s}{\dot{m}\, c_{pf}}\right). \tag{12.33}$$

Chapter 13
Steady-State Thermal Analysis

13.1 Introduction

Flow assurance in deepwater conditions is an essential part of pipeline technology for offshore oil and gas production (Harrison and Herring 2000). Flow assurance requires a carefully considered plan to insure that (1) subsea well production can be as nearly continuous as feasible during normal operation, with minimum hydrate formation and wax deposition and (2) production can be rapidly suspended when necessary and later resumed with minimum difficulty and downtime. With increasing water depth and tie-back distance, pipeline insulation has turned to be mandatory in all deepwater developments. Active heating is required when passive thermal insulation alone is not sufficient to prevent wax deposition and hydrate formation.

Two methods of active heating of deepwater pipelines have been studied in recent years: electrical heating and heating by a circulating heat medium. Norwegian oil companies, cable manufacturers, and pipeline installation companies have conducted studies on electrical heating of multiphase subsea pipelines and risers to prevent hydrate formation and wax plugs (Lervik et al. 1997, 1998; Halvorsen et al. 2000). The evaluation of technical feasibility and cost estimates have been done for a 50 Hz direct resistive heating system and for a system based on electromagnetic induction. The electrical rating of the systems depends on the heat requirement, pipe material, and the pipeline. The feasibility of the concepts has been verified through full-scale subsea tests. Results from the measurements were used to determine the characteristic parameters of the two systems on fields in the North sea.

British Petroleum (BP) planned to develop the King project in Mississippi Canyon Block 85 (MC85) via a dual well, subsea production scheme (Harrison and Herring 2000). The active heating system employs hot water circulating in the closed loop PIP (Pipe-in-Pipe) annulus. External insulation around the jacket pipe and burial/backfill provide thermal insulation of the flowlines. Heat is extracted from the exhaust waste heat of three, 3.9 MW, electrical generator turbine drivers.

C. An et al., *Structural and Thermal Analyses of Deepwater Pipes*,
https://doi.org/10.1007/978-3-030-53540-7_13

It is mandatory that passive thermal insulation be used in all deepwater flowlines for multiphase production of oil and gas as it is the most economical method to prevent hydrate formation and wax deposition. The decision as which additional flow assurance measures is suitable for a particular field development depends on engineering, economical, and environmental considerations. The electrical heating method requires additional electrical power generating capacity in the platform, for 10–30 MW depending on system configuration. No significant difficulties are expected for the mechanical design and the installation of the pipeline with electrical heating. On the other hand, the heated water circulation (HWC) system is superior from a thermodynamic point of view as discharged waste heat from the turbine drivers can be recovered without additional energetic requirement. However, the mechanical design and the installation are more expensive and involved.

In this chapter, we first present a global heat balance analysis of typical deepwater pipelines. We show that active heating is required for long pipelines due to technical limitations of passive thermal insulation systems. We then examine two methods of combined active heating and passive insulation of pipelines. For the heated water circulation in pipe-in-pipe method (HWC+PIP), we provide a closed-form analytical solution for the temperature distributions of the produced fluid and heat medium under certain simplifying hypothesis. Finally, we consider the active heating by electrical resistance on the inner steel pipe. For this case, we propose an optimum heating method to minimize electrical power requirement for a given pipeline configuration and desired minimum temperature of the produced fluid. Numerical results are shown for the last case. Significant reduction in power requirement is achieved.

13.2 Global Heat Balance Analysis

Let us consider a typical insulated pipeline for deepwater production of oil and gas. Normally, the desired thermal profile of the produced fluid and the overall heat transfer coefficient are used to specify an adequate pipeline configuration.

The overall heat transfer coefficient for a multilayered composite pipeline is given by

$$U = \frac{1}{\frac{1}{h_1} + \sum_{i=1}^{N} \frac{r_1}{k_i} \ln \frac{r_{i+1}}{r_i} + \frac{r_1}{r_{N+1} h_a}}, \tag{13.1}$$

where r_i and r_{i+1} are the inner and outer radii of the i-th layer, respectively, k_i the thermal conductivity, h_1 is the heat transfer coefficient between the innermost layer and the produced fluid flowing inside it, and h_a is the heat transfer coefficient between the outermost layer and the ambient fluid.

By the first law of thermodynamics, the global energy balance of the produced fluid can be written as

$$\dot{M}_f c_{p,f}(T_{f,in} - T_{f,out}) = U2\pi r_1 L \Delta T_m, \tag{13.2}$$

where $\dot{M}_f$ is the mass flow rate of the produced fluid, $c_{p,f}$ is the specific heat capacity, $T_{f,in}$ and $T_{f,out}$ are the inlet (wellhead) and the outlet (TLP) temperatures of the produced fluid, respectively, r_1 is the inner diameter of the innermost pipe, L is the total length of the pipeline. The average temperature difference between the produced fluid and the environmental fluid is defined as

$$\Delta T_m = \frac{(T_{f,in} - T_{f,out})}{\ln\left[(T_{f,in} - T_a)/(T_{f,out} - T_a)\right]}. \tag{13.3}$$

For a given inner radius of the pipeline r_1, a given temperature of the environmental fluid, and a desired thermal profile ($T_{f,in}$, $T_{f,out}$ and thus ΔT_m), the maximum pipeline length L_{max} that can meet the thermal profile requirement is a function of the overall heat transfer coefficient U and the heat duty of the produced fluid $\dot{M}_f c_{pi}$. From Eq. (13.2), we solve

$$L_{max} = \frac{\dot{M}_f c_{p,f}(T_{f,in} - T_{f,out})}{U2\pi r_1 \Delta T_m} = \frac{\dot{M}_f c_{p,f}}{U2\pi r_1} \ln\left[(T_{f,in} - T_a)/(T_{f,out} - T_a)\right]. \tag{13.4}$$

As the heat duty of the produced fluid, $\dot{M}_f c_{p,f}$, is usually given by the field develop plan, the maximum length of a pipeline that can meet the thermal profile requirement is inversely proportional to the overall heat transfer coefficient of the pipeline, U. The smaller the overall heat transfer coefficient, the longer the pipeline length up to which the outlet temperature of the produced fluid can meet the flow assurance requirement.

However, there are technical limitations on reducing further the overall heat transfer coefficient. For a pipeline with a length L larger than L_{max} given by Eq. (13.4), passive thermal insulation alone is not sufficient to keep the temperature of the produced fluid above a required minimum. Active heating is then required in combination with the passive thermal insulation.

If active heating is used, for a given pipeline length L, we can determine the power requirement of active heating that is necessary to meet the desired thermal profile. The global energy balance is written as, for this case

$$\dot{M}_f c_{p,f}(T_{f,in} - T_{f,out}) + \dot{Q} = U2\pi r_1 L \Delta T_m, \tag{13.5}$$

where $\dot{Q}$ stands for the total power requirement of active heating. We can solve $\dot{Q}$ from Eq. (13.5)

$$\dot{Q} = U2\pi r_1 L \Delta T_m - \dot{M}_f c_{p,f}(T_{f,in} - T_{f,out}) \tag{13.6}$$

Obviously, if all other parameters are the same, the longer the pipeline, the larger the power requirement.

The average linear heat input to the pipeline can be calculated as

$$\dot{q}_{av} = \frac{\dot{Q}}{L} = U 2\pi r_1 \Delta T_m - \frac{\dot{M}_f c_{p,f}(T_{f,in} - T_{f,out})}{L}. \tag{13.7}$$

13.3 Analysis of Heat Medium Circulation

We consider here an active heating, dual production flowline system similar to that considered by Harrison and Herring (2000). The active heating system employs heated water circulating in the closed loop pipe-in-pipe (PIP) annulus. External insulation around the jacket pipe provides passive thermal insulation of the pipeline.

For the downflow leg in which the heated water flows downward and the produced fluid flows upward, a mathematical model for one-dimensional heat transport in the produced fluid and the heated water is written as

$$\dot{M}_f c_{p,f} \frac{dT_f(x)}{dx} = -U_{fh} 2\pi r_1 L(T_f - T_h) \tag{13.8}$$

$$\dot{M}_h c_{p,h} \frac{dT_h(x)}{dx} = -U_{fh} 2\pi r_1 L(T_f - T_h) + U_{ha} 2\pi r_1 L(T_h - T_a) \tag{13.9}$$

Similarly, for the upflow leg in which both fluids flow upward, the one-dimensional heat transport equations are

$$\dot{M}_f c_{p,f} \frac{dT_f(x)}{dx} = -U_{fh} 2\pi r_1 L(T_f - T_h) \tag{13.10}$$

$$\dot{M}_h c_{p,h} \frac{dT_h(x)}{dx} = -U_{fh} 2\pi r_1 L(T_f - T_h) + U_{ha} 2\pi r_1 L(T_h - T_a), \tag{13.11}$$

where $T_h(x)$ is the temperature of the heated water, $\dot{M}_h$ its mass flow rate and $c_{p,h}$ is the specific heat capacity.

The ordinary differential equations are to be solved with two boundary conditions

$$T_f(0) = T_{f,wh} \tag{13.12}$$

$$T_h(L) = T_{h,TLP} \tag{13.13}$$

Numerical methods for boundary value problems of ordinary differential equations such as shooting method and Runge–Kutta method may be applied to solve the equations.

We present here a closed-form analytical solution of the system of ordinary differential equations, which is inspired by the works of Prof. Peter Heggs of UMIST on heat exchangers.

The downflow problem is to be solve first which consists of Eqs. (13.8) and (13.9). We introducing the temperature difference to the environmental fluid as new dependent variables

$$\theta_1 = T_f - T_a \tag{13.14}$$

$$\theta_2 = T_h - T_a \tag{13.15}$$

and rewrite the governing equations and boundary conditions in a more convenient form

$$\frac{d\theta_1}{dx} = a_{11}\theta_1 + a_{12}\theta_2 \tag{13.16}$$

$$\frac{d\theta_2}{dx} = a_{21}\theta_1 + a_{22}\theta_2 \tag{13.17}$$

$$\theta_1(0) = T_{f,wh} - T_m \tag{13.18}$$

$$\theta_2(L) = T_{h,TLP} - T_m \tag{13.19}$$

where the coefficients are defined by

$$a_{11} = -\frac{U_{fh} 2\pi r_1 L}{\dot{M}_f c_{p,f}} \tag{13.20}$$

$$a_{12} = -a_{11} \tag{13.21}$$

$$a_{21} = \frac{-U_{fh} 2\pi r_1 L}{\dot{M}_h c_{p,h}} \tag{13.22}$$

$$a_{22} = -a_{21} + \frac{U_{ha} 2\pi r_1 L}{\dot{M}_h c_{p,h}} \tag{13.23}$$

The general solution of this system of homogeneous ordinary differential equations with constant coefficients is readily obtained as

$$\theta_1(x) = c_1\xi_{11}e^{\lambda_1 x} + c_2\xi_{21}e^{\lambda_2 x} \tag{13.24}$$

$$\theta_2(x) = c_1\xi_{12}e^{\lambda_1 x} + c_2\xi_{22}e^{\lambda_2 x}, \tag{13.25}$$

where λ is the eigenvalue and ξ is the eigenvector. The constants c_1 and c_2 are determined by using the boundary conditions (13.18) and (13.19).

Once the downflow problem is solved, the upflow flow is straightforward as it is a standard initial value problem with inlet temperatures of both produced fluid and heated water.

An important application of the closed-form analytical solution is in integrated thermal–structural analysis of the pipe-in-pipe system (Solano 2001). Based on the analytical solution of the produced fluid and the heated water, the temperature fields in both pipes are readily obtained, also in analytical form.

13.4 Analysis of Direct Electrical Heating

In this section we consider the direct electrical heating of a multilayered composite pipeline. Assuming that the thermophysical properties of the produced fluid and the pipeline materials are constant, we can write the one-dimensional energy transport equation of the produced fluid as

$$\dot{M}_f c_{p,f}\frac{dT_f(x)}{dx} = -U2\pi r_1 L(T_f - T_m) + \dot{q}(x), \tag{13.26}$$

where $\dot{q}(x)$ is the linear heat input rate of direct electrical heating to the pipeline. For a given specification of the linear heat input, the temperature distribution along the pipeline can be readily obtained by solving Eq. (13.26). We are interested in determining an optimized linear heat rate distribution $\dot{q}(x)$ that minimizes the total energy consume but yet maintains the temperature of the produced fluid above a specified minimum.

Instead of formulating a constrained, continuous optimization problem which is complex to be solved, we appeal to physical insights to obtain an engineering solution of the problem. We propose only begin to heat the pipe when the temperature of the produced fluid reaches a pre-specified minimum temperature, T_{min}. The heating is to keep the produced fluid temperature at a constant temperature, this is, T_{min}. The required linear heating rate is obtained by equating the right-hand side of Eq. (13.26) to zero

$$\dot{q} = U2\pi r_1 L(T_{min} - T_m) \tag{13.27}$$

13.5 Results and Discussion

In lack of more realistic pipeline and production well data, we simulated an hypothetical pipeline with a length of 27 km and an inner diameter of 6 in.(0.1524 m). The overall heat transfer coefficient of the pipeline is 5.35 W/m^2K based on the internal pipeline diameter. The wellhead temperature of produced fluid is taken as 76.0 °C and the sea water is 4.0 °C. A mass flow rate of 14.72 kg/s is assumed with a constant density of 800.0 kg/m^3 and a specific heat of 2700.0 J/kg °C. The required minimum temperature of the produced fluid is 30.0 °C.

In Fig. 13.1 we show a comparison of the temperature distribution obtained by uniform heating of all pipeline length and the temperature distribution obtained by the optimum heating method which is our proposal. By the first method, a linear heat rate of 41.52 W/m is required to keep the temperature of produced fluid above the pre-specified minimum, T_{min} = 30.0 °C. The total power requirement is 1.12 MW. By the proposed heating method, no heat input is given for the first 15,804.7 meters of the pipeline. From that point onward, a linear heat rate of 66.60 W/m is required to keep the temperature of produced fluid above the pre-specified minimum, T_{min} = 30.0 °C. The total power requirement now is 745.6 kW, which represents a 33.4% reduction in the heating power requirement.

It can be seen clearly from Fig. 13.1 that by the first method the produced fluid is unnecessarily overheated as the area between the two curves represents the lost energy to the environment. From a thermodynamic point of view, the energy requirement is minimized if the temperature difference between the produced fluid and the sea water be minimized, which is achieved by no heating the first part of the pipeline in our method.

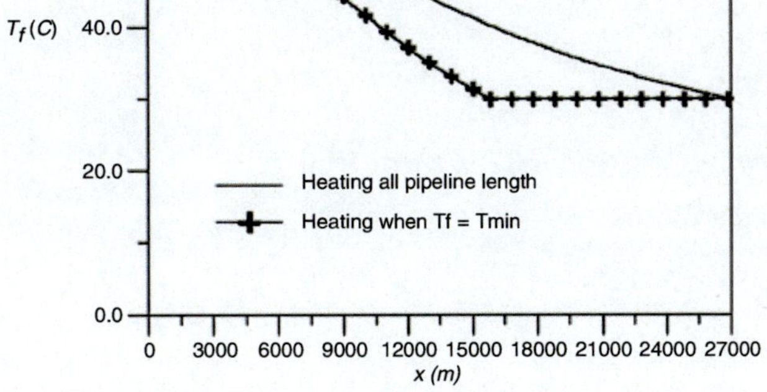

Fig. 13.1 Comparison of two heating methods

13.6 Conclusions

In this chapter, we first present a global heat balance analysis of typical deepwater pipelines for oil and gas production and show that active heating is necessary for long pipelines due to technical limitations of passive thermal insulation systems. Two methods of combined active heating and passive insulation are then considered. By the first method, active heating is provided by circulating heat medium in annulus. Under certain simplifying hypothesis, we provide a closed-form analytical solution for the temperature profiles of the produced fluid and heat medium. By the second method, active heating is provided by electrical resistance on the inner steel pipe. For this case, we propose a heating method that minimizes the power requirement for a given minimum temperature of produced fluid. Numerical results are shown for the two thermal insulation systems. For the example shown in this chapter, the proposed heating method reduces by 33.4% from the amount if uniform, all through heating is adopted. We conclude that our heating method will significantly reduce the power requirement of flow assurance by direct electrical heating.

Chapter 14
Steady-State Analysis of Heavy Oil Transportation

14.1 Introduction

The tremendous increase of oil demand and the depletion of low-viscosity oils have led to the rapid development of very large world resources of heavy oils. Heavy crudes account for a large fraction of the potentially recoverable oil reserves in the world. Production of heavy crudes is expected to increase significantly in the near future as low-viscosity crudes are being depleted. Production and transport of these heavy oils have been challenging because of their very high viscosities. The viscosities of these crudes at 25 °C vary from 1000 cP to more than 100,000 cP.

Conventional pipelining is not suitable for transporting these heavy crudes from the reservoir to the refinery because of the high viscosities involved. Generally, crude viscosities less than 200 cP are desired for pipelining. Several alternative transportation methods were proposed, including dilution with lighter crudes, injection of a water sheath around the viscous crude, or preheating the crude with subsequent heating of the pipeline. Each of these methods may have logistic, technical, or economic drawbacks for a given application. Another promising pipeline technique is the transport of viscous crudes as concentrated oil-in-water emulsions. The technical viability of this method was demonstrated in an Indonesian pipeline in 1963 and in a 13 mile long, 8 inch diameter pipeline in California.

This chapter presents the use of electrically heated pipelines for heavy oil transportation. In the past years, active electrical heating has been studied as a feasible alternative for oil and gas production in deep and ultra-deepwater, especially when associated with multilayered composite pipelines (Su 2003). Laoutr and Denntel (2001) presented a heated pipe-in-pipe system in which heating is generated by resistive cables using low voltage, which complements the insulating performance of pipe-in-pipe. Su and Cerqueira (2001) developed a mathematical model for analysis of transient heat transfer in multilayered composite pipeline during warm-up and cool-down. Su et al. (2002) presented a thermal analysis of combined active heating and passive insulation of deepwater pipelines and proposed

C. An et al., *Structural and Thermal Analyses of Deepwater Pipes*,
https://doi.org/10.1007/978-3-030-53540-7_14

a heating method that minimizes the power requirement for a given minimum temperature of produced fluid. Su et al. (2003) proposed a new concept to combine thermal insulation and active heating by inserting strips of electrical resistance into the sandwich pipes. Initially a global heat balance analysis to determine the energy input requirement was carried out and then a steady-state thermal analysis to determine the temperature distribution in a cross of the sandwich pipes under typical production conditions in ultra-deepwater. A mathematical model was developed for the analysis of steady-state heat transfer in the sandwich pipes. The analysis of transient heat transfer in the sandwich pipelines with active electrical heating was presented by Cerqueira et al. (2004). The mathematical model governing the heat conduction in the composite pipeline and the energy transport in the produced fluid was solved by using finite difference methods. As unplanned cool-down of the pipelines is most critical to safe and economical operation of pipelines in deep and ultra-deepwater conditions, numerical results of computational simulation of cool-down for three sandwich pipeline configurations under typical production conditions were also presented.

In this chapter, we present a simplified thermal-hydraulic analysis of heated sandwich pipelines for heavy oil transportation. The temperature distribution of the produced fluid is determined by solving the energy transport equation for different heating rates. The pressure drop across the pipeline length is then determined by integrating the local pressure gradient as a function of the local temperature of the produced fluid. We show through a numerical example that active heating can greatly enhance the production rate of a pipeline for a given pressure drop.

14.2 Analysis

We consider the transport of a heavy oil in a multilayer composite pipeline with direct electrical heating. The thermodynamic and transport properties of the produced fluid are specified through the density ρ_f, the thermal conductivity $k(T)$, the specific heat $c_{p,f}(T)$, and the viscosity $\mu(T)$. The geometrical configuration and material properties of the multilayered composite pipeline are specified. The composite medium consists of n concentrically cylindrical layers, as shown in Fig. 12.2.

Each layer is assumed to be homogeneous, isotropic, and with constant thermal properties. The adjacent layers are assumed to be in perfect thermal contact. The overall heat transfer coefficient for a multilayered composite pipeline is defined as

$$U = \frac{1}{\frac{1}{h_1} + \sum_{i=1}^{N} \frac{r_1}{k_i} \ln \frac{r_{i+1}}{r_i} + \frac{r_1}{r_{N+1} h_a}}, \tag{14.1}$$

where r_i and r_{i+1} are the inner and outer radii of the i-th layer, respectively, k_i is the thermal conductivity, h_1 is the heat transfer coefficient between the innermost layer and the produced fluid flowing inside it, and h_a is the heat transfer coefficient between the outermost layer and the environmental fluid.

The one-dimensional energy transport equation of the produced fluid is written as

$$\dot{M}_f c_{p,f}(T)\frac{dT_f(z)}{dz} = -U2\pi r_1(T_f - T_m) + \dot{q}(z), \quad 0 < x < L \tag{14.2}$$

subject to the initial condition

$$T_f(0) = T_{f,in}, \tag{14.3}$$

where $\dot{M}_f$ is the mass flow rate of the produced fluid, $c_{p,f}$ is the specific heat, $T_{f,in}$ is the inlet (wellhead) temperature of the produced fluid, r_1 is the inner diameter of the innermost steel pipe, L is the length of the pipeline, and $\dot{q}(z)$ is the linear heat input rate of direct electrical heating to the pipeline. For a given specification of the linear heat input, the temperature distribution along the pipeline can be readily obtained by solving Eq. (14.2).

The pressure drop across the pipeline length Δp is calculated by integrating the local pressure gradient

$$\Delta p = \int_{z=0}^{z=L} -\left(\frac{dp}{dz}\right)dz. \tag{14.4}$$

The local pressure gradient $-dp/dz$ is determined through the Fanning friction factor

$$-\frac{dp}{dz} = \frac{f}{D}\frac{1}{2}\rho_f V_m^2, \tag{14.5}$$

where D is the inner diameter of the pipeline and V_m is the mean velocity of the produced fluid.

The friction factor is calculated by using a recently published correlation (Garcia et al. 2003)

$$f = 0.0925\text{Re}^{-0.2534} + \frac{13.98\text{Re}^{-0.9501} - 0.0925\text{Re}^{-0.2534}}{\left(1 + \left(\frac{\text{Re}}{293}\right)^{4.864}\right)^{0.1972}}, \tag{14.6}$$

where the Reynolds number is defined as

$$\text{Re} = \frac{\rho_f V_m D}{\mu}. \tag{14.7}$$

14.3 Results and Discussion

Due to the lack of more realistic pipeline and production well data, hypothetical conditions have been applied for sandwich pipelines with inner diameter of 6 in (0.1524 m). The pipelines are in sandwich pipe configuration with internal and external steel pipes and with polypropylene as thermal insulation material in the annulus. The steel pipes are of 1/8 in (0.003725 m) thickness and the thermal insulation is of 3 in (0.0762 m) thickness. The length of the pipeline is 25 km. The thermophysical properties of the pipelines are given in Table 14.1. The wellhead temperature of produced fluid is taken as 90 °C and the sea water as 4.0 °C. A constant density of 800 kg/m^3 and a specific heat of 2700 J/kg °C are assumed.

The viscosity of a typical heavy oil (°API = 18) is given by the following empirical expression obtained by data-fitting

$$\mu_0 = 7094.54e^{-0.05878T}. \tag{14.8}$$

The variation of the viscosity as function of temperature is shown in Fig. 14.1.

The temperature distributions of the produced fluid along the pipeline are shown in Fig. 14.2 for a mass flow rate of 20.0 kg/s for three different linear heat input rates. For the pipeline of a length of 25 km, these values correspond to the total power input of 0, 500 kW, and 1 MW respectively. With the same inlet temperature of 90 °C, the outline temperatures are 45, 51, and 58 °C for the three heating rates. The different thermal profiles will result in different pressure drop across the pipeline due to the dependence of the viscosity in temperature.

The dependence of the pressure drop in heating rate is shown in Fig. 14.4 for mass flow rates of 20, 25, and 30 kg/s. Figure 14.3 is to be used for situations where

Table 14.1 Thermophysical properties of the composite pipelines

Material	ρ (kg/m^3)	c_p (J/kg °C)	k (W/m °C)
Stainless steel	7850	486	54
Polypropylene	775	2000	0.17

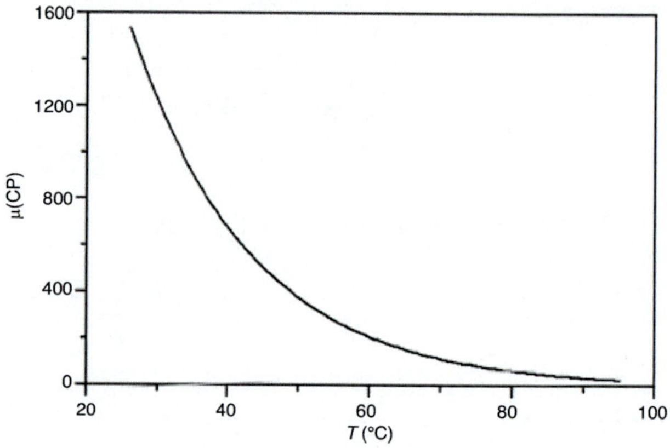

Fig. 14.1 The viscosity of a typical heavy oil as a function of temperature

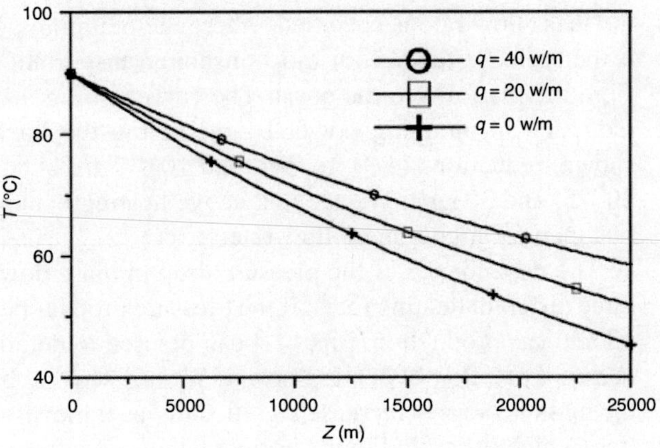

Fig. 14.2 Temperature distribution of the produced fluid along the pipeline for different linear heating rate: $\dot{M} = 20\,\text{kg/s}$

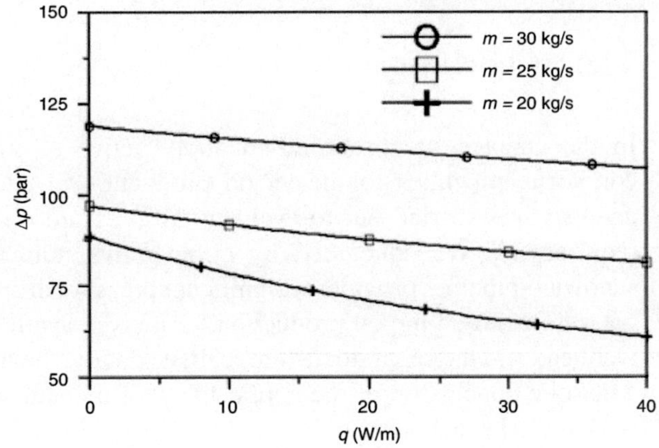

Fig. 14.3 Pressure drop of the produced fluid as a function of mass flow rate for different linear heating rates

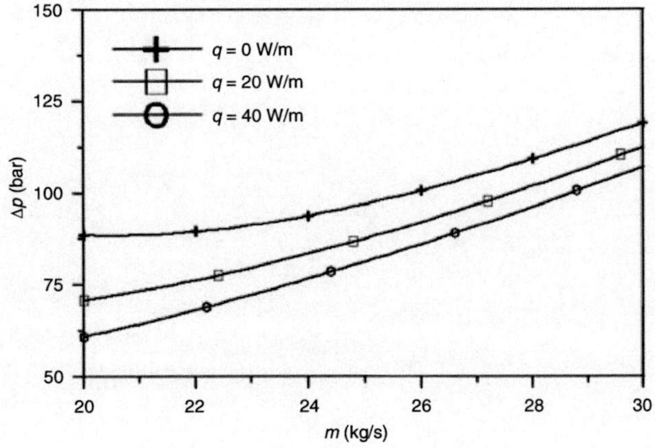

Fig. 14.4 Pressure drop of the produced fluid as a function of linear heating rate for different mass flow rates

the mass flow rate is specified, where the pumping power is to be determined. This is the typical situation for most onshore transportation and for direct transportation from the platform to the coast. The curves in Fig. 14.3 can be used to estimate the reduction in pumping power by increasing the linear heating rate. For the cases shown, reductions of 31.1, 16.1, and 10.0% are obtained for the mass flow rates of 20, 25, and 30 kg/s. We see that active heating is more crucial for lower mass flow rate than for higher mass flow rate.

The dependence of the pressure drop in mass flow rate is shown in Fig. 14.4 for three different heating rates. If the pressure drop is specified as the case of deepwater oil and gas production, Fig. 14.4 can be used to determine the enhanced production rate as a function of heating power. We can see that by increasing the heating rates, the mass flow rates increase significantly as it moves along a constant pressure drop line, which is parallel to the horizontal axis.

14.4 Conclusions

In the chapter, we proposed to apply active electrical heating in multilayered composite pipelines for heavy oil production in deepwater. A thermal-hydraulic analysis was carried out to evaluate the pressure drop reduction and power input requirement. We concluded by comparative numerical studies that the heated sandwich pipeline provides a significant pressure drop reduction and thus a boost of mass flow rate of the oil production for a given available pressure drop between the wellhead and the separator. More realistic data of the fluid properties are required for a definite conclusion of the applicability of the actively heated sandwich pipelines in heavy oil production.

Chapter 15
Analysis of Direct Electrical Heating

15.1 Introduction

Flow assurance in deepwater conditions is an essential part of pipeline technology for offshore oil and gas production (Su 2003). Flow assurance requires a carefully considered plan to insure that (1) subsea well production can be as nearly continuous as feasible during normal operation, with minimum hydrate formation and wax deposition, (2) production also can be rapidly suspended when necessary and later resumed with minimum difficulty and downtime. With increasing water depth and tie-back distance, pipeline insulation has turned to be mandatory in all deepwater developments. Active heating is required when passive thermal insulation alone is not sufficient to prevent wax deposition and hydrate formation.

Two methods of active heating of deepwater pipelines have been studied in recent years: electrical heating and heating by a circulating heat medium. Norwegian oil companies, cable manufacturers, and pipeline installation companies have conducted studies on electrical heating of multiphase subsea pipelines and risers to prevent hydrate formation and wax plugs (Lervik et al. 1997, 1998; Halvorsen et al. 2000). The evaluation of technical feasibility and cost estimates have been done for a 50 Hz direct resistive heating system and for a system based on electromagnetic induction. The electrical rating of the systems depends on the heat requirement, pipe material, and the pipeline. The feasibility of the concepts has been verified through full-scale subsea tests. Results from the measurements were used to determine the characteristic parameters of the two systems on fields in the North sea. British Petroleum (BP) planned to develop the King project in Mississippi Canyon Block 85 (MC85) via a dual well, subsea production scheme (Harrison and Herring 2000). The active heating system employs hot water circulating in the closed loop PIP (Pipe-in-Pipe) annulus. External insulation around the jacket pipe and burial/backfill provide thermal insulation of the flowlines. Heat is extracted from the exhaust waste heat of three, 3.9 MW, electrical generator turbine drivers.

C. An et al., *Structural and Thermal Analyses of Deepwater Pipes*,
https://doi.org/10.1007/978-3-030-53540-7_15

It is mandatory that passive thermal insulation is used in all deepwater flowlines for multiphase production of oil and gas as it is the most economical method to prevent hydrate formation and wax deposition. Pipe-in-pipe systems have been lately proposed as viable solutions to such problems (Nutall and Rogers 1998; Franklin and Wright 1999; Trout and Sahota 1999). They consist of two concentric metal pipes in which the annulus is either filled with a non-structural insulating material or is used to carry water for well injection, umbilical cables, etc. Usually, internal and external pipes are designed independently against failure under internal and external pressures, respectively, combined with installation loads, mainly longitudinal bending. The factors governing the collapse and propagation of buckles in single pipes and pipe-in-pipe systems under external pressure have been extensively studied in the past, so that, nowadays deepwater pipes can be safely designed (Murphey and Langner 1985; Yeh and Kyriakides 1988; Kyriakides 2002; Kyriakides and Vogler 2002; Estefen et al. 2005a). Laoutr and Denntel (2001) presented a heated pipe-in-pipe system in which heat is generated by resistive cables using low voltage, which complements the insulating performance of pipe-in-pipe. Su and Cerqueira (2001) developed a mathematical model for analysis of transient heat transfer in multilayered composite pipeline during warm-up and cool-down. Su et al. (2002) presented a thermal analysis of combined active heating and passive insulation of deepwater pipelines and proposed a heating method that minimizes the power requirement for a given minimum temperature of produced fluid.

In this chapter, we present a new concept to combine thermal insulation and active heating by inserting strips of electrical resistance into the sandwich pipes. We first carry out a global heat balance analysis to determine the energy input requirement and then a steady-state thermal analysis to determine the temperature distribution in a cross section of the sandwich pipes under typical production conditions in ultra-deepwater. A mathematical model is developed for the analysis of steady-state heat transfer in the sandwich pipes.

15.2 Global Heat Balance Analysis

Let us consider a typical insulated pipeline with active electrical heating for deepwater production of oil and gas. Normally, the desired thermal profile of the produced fluid and the overall heat transfer coefficient are used to specify an adequate pipeline configuration.

The overall heat transfer coefficient for a multilayered composite pipeline is defined as

$$U = \frac{1}{\frac{1}{h_1} + \sum_{i=1}^{N} \frac{r_1}{k_i} \ln \frac{r_{i+1}}{r_i} + \frac{r_1}{r_{N+1} h_a}}, \tag{15.1}$$

where r_i and r_{i+1} are the inner and outer radii of the i-th layer, respectively, k_i is the thermal conductivity, h_1 is the heat transfer coefficient between the innermost layer and the produced fluid flowing inside it, and h_a is the heat transfer coefficient between the outermost layer and the environmental fluid.

By the first law of thermodynamics, the global energy balance of the produced fluid can be written as

$$\dot{M}_f c_{p,f}(T_{f,in} - T_{f,out}) + \dot{Q} = U 2\pi r_1 L \Delta T_m, \tag{15.2}$$

where $\dot{M}_f$ is the mass flow rate of the produced fluid, $c_{p,f}$ is the specific heat capacity, $T_{f,in}$ and $T_{f,out}$ are the inlet (wellhead) and the outlet (TLP) temperatures of the produced fluid respectively, r_1 is the inner diameter of the innermost steel pipe, L is the length of the pipeline, and $\dot{Q}$ stands for the power requirement of active heating. We can solve $\dot{Q}$ from Eq. (15.2)

$$\dot{Q} = U 2\pi r_1 L \Delta T_m - \dot{M}_f c_{p,f}(T_{f,in} - T_{f,out}). \tag{15.3}$$

We consider the direct electrical heating of a multilayered composite pipeline. Assuming that the thermophysical properties of the produced fluid and the pipeline materials are constant, we can write the one-dimensional energy transport equation of the produced fluid as

$$\dot{M}_f c_{p,f} \frac{dT_f(x)}{dx} = -U 2\pi r_1 L (T_f - T_m) + \dot{q}(x), \tag{15.4}$$

where $\dot{q}(x)$ is the linear heat input rate of direct electrical heating to the pipeline. For a given specification of the linear heat input, the temperature distribution along the pipeline can be readily obtained by solving Eq. (15.4).

15.3 Physical Problem

The physical problem consists of a sandwich pipe composed of two concentric metal pipes with insulation material in the annulus. We consider that each layer is homogeneous, isotropic, and with constant thermal properties. It is also considered that adjacent layers are in perfect thermal contact. Four strips of electrical heater are located symmetrically over the outer surface of the inner metal pipe. Given the properties of each layer, its dimension and the power of the heat generation, our problem is to predict the temperature distribution in these layers (Fig. 15.1).

15.4 Electrical Heating

Studies have been done about the electrical heating as a way to prevent wax deposition and hydrate formation. The active heating system is an alternative

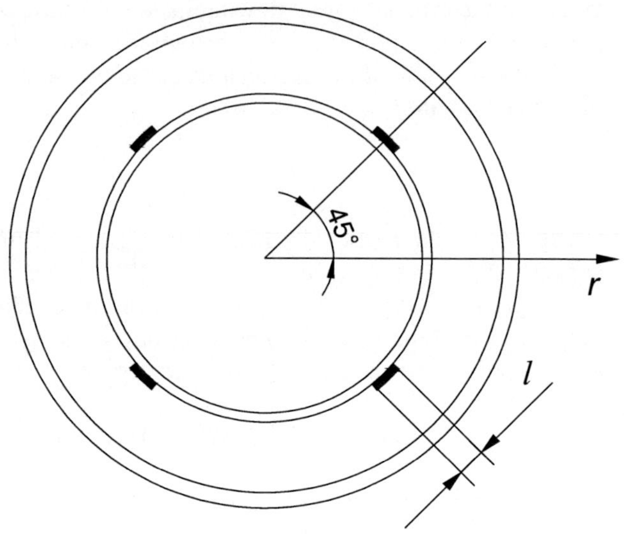

Fig. 15.1 Cross section of a multilayered composite pipeline

solution when passive insulation is not enough to prevent these depositions in the flowline. In steady-state conditions, passive insulation alone may attend flow assurance requirements, but during transient events such as warm-up and cool-down or in exceptional situations such as plants shutdowns curative solutions are often used. These methods can, sometimes become complicated and expensive.

The heating system is basically composed by copper cables grouped in sets distributed around the flowline. The heat input is produced by the electrical activation of the resistive heating cables.

15.5 Mathematical Modeling

We analyze here only 1/8 of the cross section of the pipe, given the symmetry of the problem. The mathematical formulation of the bi-dimensional heat conduction problem is written as

$$\frac{k_i}{r}\left(r\frac{\partial T_i}{\partial r}\right) + \frac{k_i}{r^2}\frac{\partial^2 T_i}{\partial \theta^2} + g_i = 0 \tag{15.5}$$

$$r_1 < r < r_2$$

$$\frac{k_1}{r}\left(r\frac{\partial T_1}{\partial r}\right) + \frac{k_1}{r^2}\frac{\partial^2 T_1}{\partial \theta^2} = 0 \tag{15.6}$$

$$r_1 < r < r_2$$

$$\frac{k_2}{r}\left(r\frac{\partial T_2}{\partial r}\right)+\frac{k_2}{r^2}\frac{\partial^2 T_2}{\partial \theta^2}+g_2=0 \tag{15.7}$$

$$r_2<r<r_3,\quad o<\theta<\frac{\pi}{2}-\Phi,\quad \frac{\pi}{2}-\Phi<\theta<\frac{\pi}{2}$$

$$\frac{k_3}{r}\left(r\frac{\partial T_3}{\partial r}\right)+\frac{k_3}{r^2}\frac{\partial^2 T_3}{\partial \theta^2}+g_3=0 \tag{15.8}$$

$$r_3<r<r_4$$

$$\frac{k_H}{r}\left(r\frac{\partial T_H}{\partial r}\right)+\frac{k_H}{r^2}\frac{\partial^2 T_H}{\partial \theta^2}+g_H=0 \tag{15.9}$$

$$r_2<r<r_2+\delta,\quad \frac{\pi}{2}-\phi<\theta<\frac{\pi}{2},$$

where T is the temperature, k the thermal conductivity. The subscripts 1, 2, and 3 indicate each layer and the subscripts H indicate the heater. This problem is solved with the following boundary conditions in A and B Fig. 15.2:

$$\frac{\partial T}{\partial \theta}=0,\theta=0,\theta=\frac{\pi}{2} \tag{15.10}$$

and the following convective conditions:

$$k_1\frac{\partial T}{\partial r}=h_1(T_1-T_f),r=r_1 \tag{15.11}$$

$$k_3\frac{\partial T}{\partial r}=h_a(T_3-T_m),r=r_4, \tag{15.12}$$

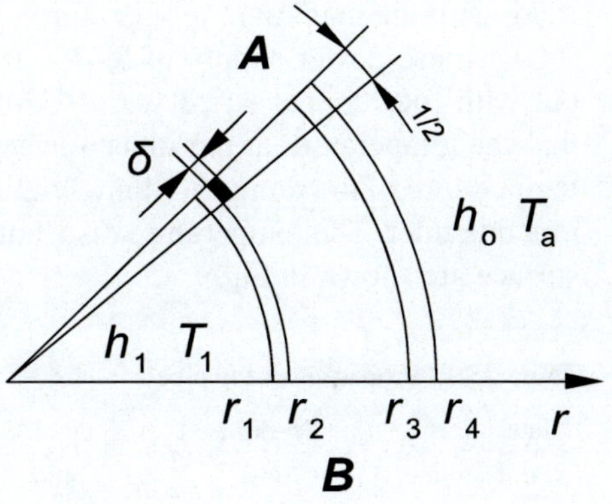

Fig. 15.2 Longitudinal view of a pipeline

where h_1 is the heat transfer coefficient of between innermost layer and the fluid inside it, T_f is the temperature of the fluid transported, h_a is the heat transfer coefficient of outermost surface and the environmental fluid, and T_1 is the temperature of environmental fluid.

15.6 Results and Discussion

Due to the lack of more realistic pipeline and production well data, we simulated an hypothetical pipeline with a length of 27 km and an inner diameter of 6 in. (0.1524 m). The overall heat transfer coefficient of the pipeline is 5.35 W/m^{2}K based on the internal pipeline diameter. The wellhead temperature of produced fluid is taken as 76 °C and the sea water as 4.0 °C. A mass flow rate of 14.72 kg/s is assumed with a constant density of 800 kg/m^3 and a specific heat of 2700 J/kg °C. The required minimum temperature of the produced fluid is 30 °C. The thermophysical properties of the pipeline are given in Table 15.1.

A linear heat rate of 72 W/m is required to keep the temperature of produced fluid above the pre-specified minimum, T_{min} = 39 °C, if uniform heating of all pipeline length is applied. The total power requirement is 1.944 MW. We first consider a supply voltage of 5.4 kV for the 27 km pipeline. Four copper strips are inserted between the inner steel pipe and the thermal insulation material, each with 1 mm thickness and 8 mm width. The temperature distributions of the produced fluid and of inner surface of the pipe are shown in Fig. 15.3. We can see that the temperature at the inner surface of the inner steel pipe is 3 °C above the temperature of the produced fluid, with the maximum temperature of 78.18 °C at the pipeline inlet.

With the same uniform heating method but with a higher supply voltage of 8.1 kV for the 27 km pipeline, four smaller copper strips are required, with 1 mm thickness and 3.5 mm width each. The temperature distributions of the produced fluid and of pipe inner surface are shown in Fig. 15.4. We can see that the temperature at the inner surface of the inner steel pipe are 6 °C above the temperature of the produced fluid, with the maximum temperature of 81.54 °C at the pipeline inlet.

If we use power supply of 5.4 kV for the 27 km pipeline with uniform heating, but with four copper strips with 0.5 mm thickness and 16 mm width. We can see that the temperature at the inner surface of the inner steel pipe are 1 °C above the temperature of the produced fluid, with the maximum temperature of 76.86 °C at the pipeline inlet. The temperature distributions of the produced fluid and of pipe inner surface are shown in Fig. 15.5.

Table 15.1 Properties of the pipe's layers

Material	ρ (kg/m^3)	c_p (J/kg °C)	k (W/m °C)	Thickness (mm)
Stainless steel	7850	486	54	3.175
Polypropylene	775	2000	0.17	76.2
Stainless steel	7850	486	54	3.175
Copper	8900.0	380	380	

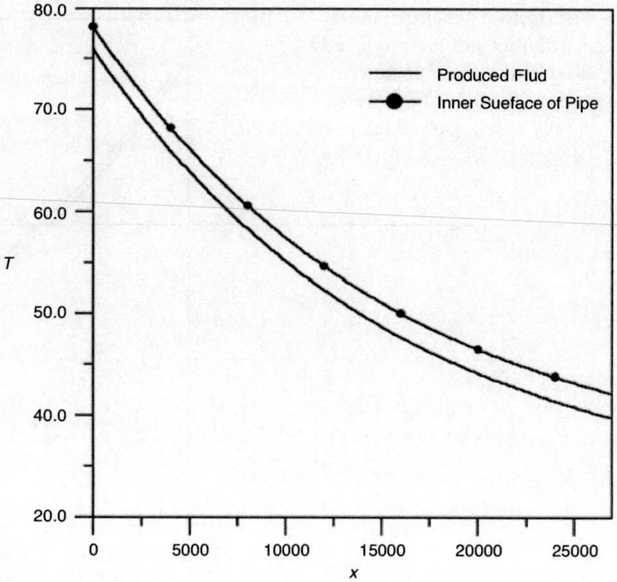

Fig. 15.3 The temperature distributions of the produced fluid and of the pipe inner surface for 0.2 V/m power supply

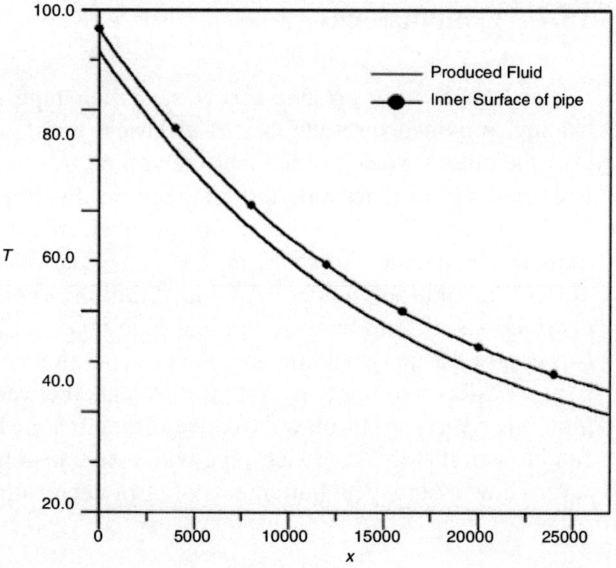

Fig. 15.4 The temperature distributions of the produced fluid and of the pipe inner surface for 0.3 V/m power supply

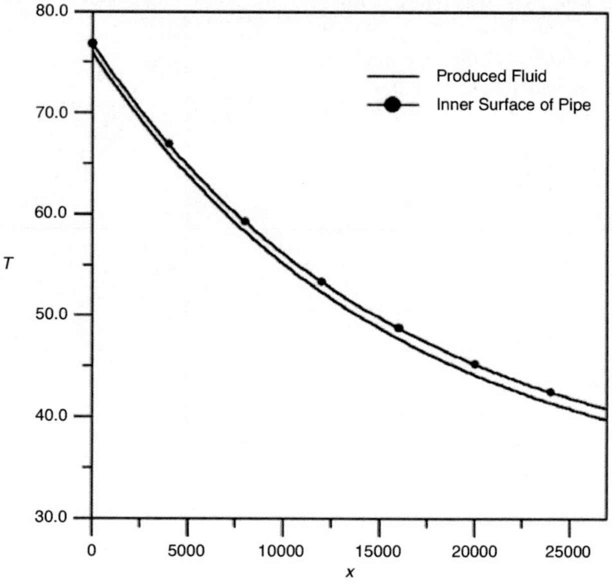

Fig. 15.5 The temperature distributions of the produced fluid and of the pipe inner surface for 0.2 V/m power supply and 0.5 mm thick heater

15.7 Conclusions

In this chapter, we present a new sandwich pipe system with active electrical heating, provided by a number of electrical resistance strips stuck longitudinally over the outer surface of the inner metal pipe. A steady-state heat transfer analysis is carried out to determine the temperature distribution in a cross section of the sandwich pipe, the power requirement of electrical heating, and the overall heat transfer coefficient. We see from the numerical results that with a power supply of 0.3 V/m and four copper strip with 1 mm thickness and 4 mm width, acceptable distributions of fluid temperature and pipe inner surface temperature are obtained. Lower power supply will increase the cross section area of the heater; thinner heater strip will decrease the temperature difference between the produced fluid and the pipe inner surface but may introduce difficulties in fabrication and installation. It can be seen that the sandwich pipe with active heating is a viable solution to meet severe flow assurance requirements of ultra-deepwater oil production.

Chapter 16
Transient Analysis of Multilayer Composite Pipelines with Active Heating

16.1 Introduction

The analysis of transient heat transfer in multilayer composite pipelines with active heating is an essential aspect for flow assurance design and operating strategies of deepwater subsea pipelines, such as pipe-in-pipe or sandwich systems. With increasing water depth and tie-back distance, pipeline insulation has turned to be mandatory in all types of deepwater developments. For subsea production pipelines, chemical inhibitors or active heating is required when passive thermal insulation alone is not sufficient to prevent wax deposition and hydrate formation probably appeared during warm-up or cool-down periods (Sarmento et al. 2004; Nysveen et al. 2007; Basogul and Kecebas 2011). Accurate analysis of transient heat transfer in composite pipeline is necessary to predict temperature evolution along the pipeline as to determine necessary quantity of chemical inhibitors or suitable means of heating. As an example, Su et al. (2005) solved the mathematical models governing the heat conduction in the composite pipeline and the energy transport in the produced fluid by using finite difference methods.

Transient heat transfer in multilayered composite plates, cylinders, and spheres is of great interest in a number of engineering applications (An and Su 2011; Maestre et al. 2010; Price and Smith 1995; Regis et al. 2000; Su and Cotta 2001; Su et al. 2005; Cha et al. 1995; de Monte 2002; Jain et al. 2010). The use of composite media is necessary when the thermal and mechanical properties of a single layer are not sufficient as to fulfill both thermal and mechanical requirements. Various methods are available for the determination of the transient temperature distribution in multilayered composite media, such as the Laplace transform method (Li and Lai 2013), the orthogonal expansion technique (Mikhailov et al. 1983), the Green's function approach (Haji-Sheikh and Beck 1990), and the finite integral transform technique (Mikhailov and Özişik 1984).

Recently, heat conduction in cylindrical composite laminates has attracted much attention due to its relevance to thermal performance of composite pipes and vessels.

C. An et al., *Structural and Thermal Analyses of Deepwater Pipes*,
https://doi.org/10.1007/978-3-030-53540-7_16

Using the Sturm–Liouville theorem to derive an appropriate Fourier transformation, Kayhani et al. (2012) presented a general analytical solution for conductive heat transfer in cylindrical composite laminates with general boundary conditions which are combinations of conduction, convection, and radiation. Extending previous work, Delouei et al. (2012) applied Laplace transformation to change the domain of the solutions from time into the frequency, and performed the inverse Laplace transformation based on the Meromorphic function method to find the transient temperature distribution in cylindrical composite laminates. Chen et al. (2003) introduced a frequency-domain regression method to calculate the thermal response factors and conduction transfer function coefficients of multilayer cylindrical walls. Li and Lai (2013) derived a set of classical explicit analytical solutions for a two-layer composite hollow cylinder medium with general inhomogeneous boundary conditions. Based on the laminate approximation theory, Wang (2013) performed the transient thermal analysis in functionally graded cylindrical structures, the analytical solution of which was obtained by the state space method and the initial parameter method. Wang and Liu (2013) employed the method of separation of variables to develop an analytical solution for two-dimensional transient heat conduction in a fiber reinforced multilayer cylinder composites.

From an engineering point of view, most of the above-mentioned methods (such as the Laplace transform method, the orthogonal expansion technique, the Green's function approach, etc.) are not convenient to use in practice because of the complicated procedures involved in the analytical derivation and the numerical computation. When the exact temperature distribution in the composite pipelines is not of primary interest, the lumped parameter approach can be adopted which has been widely used in a variety of thermal engineering applications. The classical lumped parameter approach is in general restricted to problems with low to moderate temperature gradients, typically with Biot number less than 0.1. In most engineering problems, the Biot number is much higher. Cotta and Mikhailov (1997) proposed a systematic formalism to provide improved lumped parameter formulations for steady and transient heat conduction problems based on Hermite approximations for integrals that define averaged temperature and fluxes. Regis et al. (2000) developed an improved lumped analysis of transient heat conduction in a nuclear fuel rod which was represented by a two-region concentric cylinder. An and Su (2013) presented a lumped parameter model for one-dimensional heat conduction with melting of a phase change material (PCM) slab with volumetric heat generation. Naveira et al. (2009) presented a hybrid numerical–analytical solution for transient laminar forced convection over flat plates of non-negligible thickness, where an improved lumped-differential formulation for the transversally averaged wall temperature was applied.

In this chapter, we applied improved lumped-differential formulations based on two-points Hermite approximations for integrals for the analysis of transient heat conduction in multilayered composite pipelines with active heating. The improved lumped models governing the heat conduction in the composite pipeline and the transient energy equation for the produced fluid are solved by using finite difference methods. With the proposed method, we analyze the transient heat transfer in stainless steel–polypropylene–stainless steel sandwich pipes (SP) with active electrical heating. Convergence behaviors of the average temperature of each layer and the bulk temperature of the produced fluid calculated by using the improved lumped models ($H_{0,0}/H_{1,1}$ and $H_{0,0}/H_{1,1}$ approximations) against the number of grid points along the length direction of pipelines are presented. In addition, the effects of the linear heat generation rate and the average velocity on the bulk temperature distribution of the produced fluid are investigated.

16.2 The Mathematical Formulation

Heat Conduction in Composite Pipeline Consider a multilayered composite pipeline consisting of N concentrically cylindrical layers, as shown in Fig. 16.1a, with r and z being the coordinates in the radial and axial directions, respectively.

Each layer is assumed to be homogeneous, isotropic, and with constant thermal properties. The adjacent layers are assumed to be in perfect thermal contact. The heating system is basically composed by power cables distributed around the flowline. The heat is generated as electric current passing through the cables. The heat conduction is assumed to be axisymmetric. The mathematical formulation of the transient heat conduction problem is written as

$$\frac{\partial T_i(r,z,t)}{\partial t} = \frac{k_i}{\rho_i c_{pi} r}\frac{\partial}{\partial r}\left(r\frac{\partial T_i(r,z,t)}{\partial r}\right) + \frac{g_i(r,z,t)}{\rho_i c_{pi}}, \quad r_i < r < r_{i+1},$$
$$i = 1, 2, \ldots, N, \tag{16.1}$$

where $T_i(r, z, t)$ is the temperature in the i-th layer, t the time, k_i the thermal conductivity, ρ_i the density, c_{pi} the specific heat, and g_i the volumetric heat generation rate. The inner and outer radii of the i-th layer are r_i and r_{i+1}, respectively. Equation (16.1) is to be solved with the following boundary and interface conditions:

$$-k_1\frac{\partial T_1(r,z,t)}{\partial r} = h_1(T_f(z,t) - T_1(r,z,t)), \quad \text{at} \quad r = r_1, \quad \text{and} \quad 0 < z < L. \tag{16.2}$$

$$-k_N\frac{\partial T_N(r,z,t)}{\partial r} = h_2(T_N(r,z,t) - T_m), \quad \text{at} \quad r = r_{N+1}, \quad \text{and} \quad 0 < z < L. \tag{16.3}$$

$$T_i(r,z,t) = T_{i+1}(r,z,t), \quad \text{at} \quad r = r_{i+1}, \quad \text{and} \quad 0 < z < L.,$$
$$i = 1, 2, \ldots, N-1, \tag{16.4}$$

$$k_i\frac{\partial T_i(r,z,t)}{\partial r} = k_{i+1}\frac{\partial T_{i+1}(r,z,t)}{\partial r}, \quad \text{at} \quad r = r_{i+1}, \quad i = 1, 2, \ldots, N-1, \tag{16.5}$$

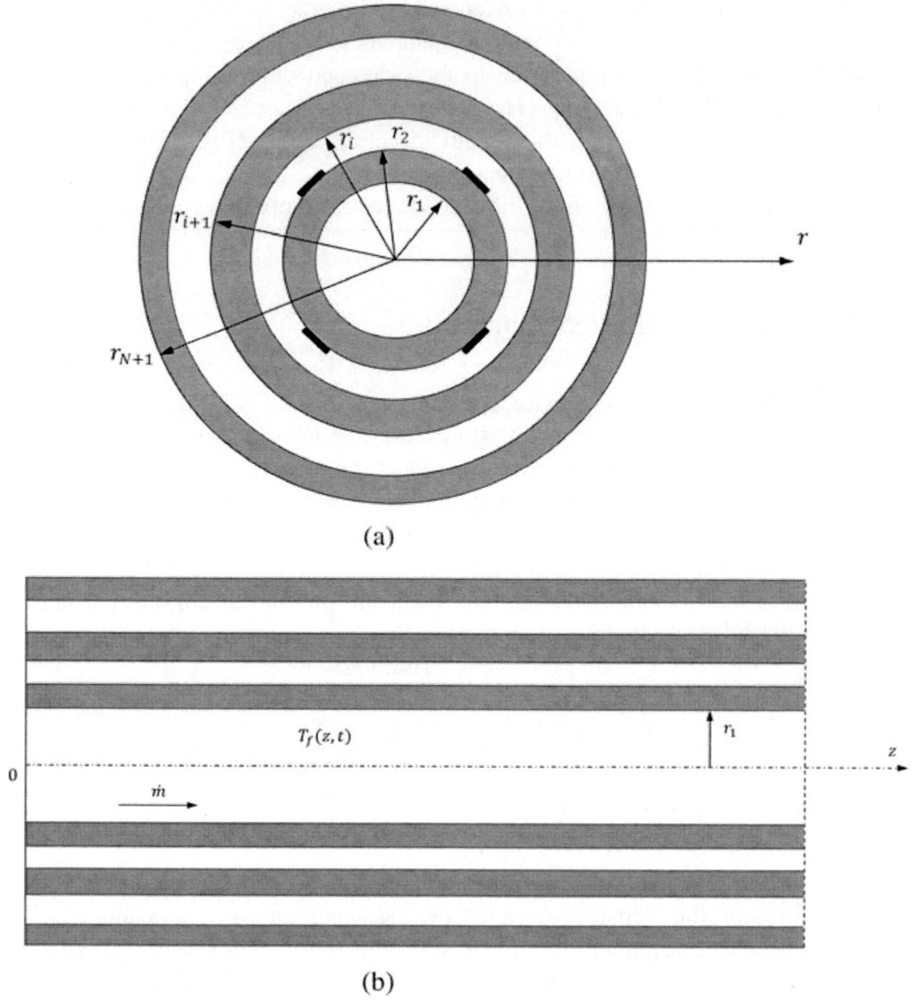

Fig. 16.1 Illustration of a multilayered composite pipeline. (**a**) Cross section. (**b**) Longitudinal view

where T_f is the temperature of the fluid transported in the pipeline, h_1 the heat transfer coefficient between the innermost layer and the internal fluid, h_2 the heat transfer coefficient between the outermost surface and the environmental fluid, and T_m the temperature of the environmental fluid. The initial condition and an inlet boundary condition in each layer are

$$T_i(r, z, 0) = T_{i0}(r, z), \quad r_i < r < r_{i+1}, \quad i = 1, 2, \ldots, N. \tag{16.6}$$

$$T_i(r, 0, t) = T_{fin}, \quad \text{at} \quad z = 0. \tag{16.7}$$

Energy Transport in Produced Fluid Neglecting the effects of flow transient, we consider a steady, fully developed flow with an average velocity U_f of a produced fluid with constant properties in a pipeline of circular transversal section, as shown in Fig. 16.1b. The one-dimensional transient energy equation for the produced fluid is written as

$$\rho_f c_{pf}\frac{\partial T_f(z,t)}{\partial t} + \rho_f c_{pf} U_f \frac{\partial T_f(z,t)}{\partial z} = \frac{h_1 P_w}{A_f}(T_1(r,z,t)|_{r=r_1} - T_f(z,t)),$$
$$\text{in} \quad 0 < z < L \quad \text{and} \quad t > 0 \qquad (16.8)$$

where $T_f(z,t)$ is the bulk temperature of the produced fluid as a function of spatial coordinate z and time t, ρ_f the density of the produced fluid, c_{pf} the specific heat of the fluid, P_w the inner perimeter of the pipeline, and A_f the cross section area of the flow passage. Equation (16.8) is to be solved with an initial temperature distribution of the fluid along the length of pipeline and an inlet boundary condition in the wellhead ($z = 0$)

$$T_f(z,0) = T_{f0}(z), \quad \text{at} \quad t = 0, \qquad (16.9)$$

$$T_f(0,t) = T_{fin}, \quad \text{at} \quad z = 0. \qquad (16.10)$$

16.3 Improved Lumped Models

Let us introduce the spatially averaged dimensionless temperature as follows:

$$T_{av,i}(z,t) = \frac{2}{r_{i+1}^2 - r_i^2}\int_{r_i}^{r_{i+1}} rT_i(r,z,t)\mathrm{d}r. \qquad (16.11)$$

Operate Eqs. (16.1) and (16.6) by $\frac{2}{r_{i+1}^2 - r_i^2}\int_{r_i}^{r_{i+1}} r\mathrm{d}r$ and using the definition of average temperature, Eq. (16.11), we get

$$\frac{\partial T_{av,i}(z,t)}{\partial t} = \frac{2k_i}{(r_{i+1}^2 - r_i^2)\rho_i c_{pi}}\left(r_{i+1}\frac{\partial T_i}{\partial r}\bigg|_{r=r_{i+1}} - r_i \frac{\partial T_i}{\partial r}\bigg|_{r=r_i} \right)$$
$$+ \frac{g_i(z,t)}{\rho_i c_{pi}}, \quad \text{in} \quad 0 < z < L \quad \text{and} \quad t > 0, \quad i = 1, 2, \ldots, N, \qquad (16.12)$$

and

$$T_{av,i}(z,0) = \frac{2}{r_{i+1}^2 - r_i^2} \int_{r_i}^{r_{i+1}} r T_{i0}(r,z) \mathrm{d}r, \quad i = 1, 2, \ldots, N, \tag{16.13}$$

when considering the volumetric heat generation rate is independent of the radial coordinate, viz., $g_i(r,z,t) = g_i(z,t)$. Equation (16.12) is an equivalent integro-differential formulation of the original mathematical model Eq. (16.1), with no approximation involved. The basic idea of improved lumped-differential formulations is to provide better relations between the boundary temperatures and boundary heat fluxes than that used by the classical lumped system analysis (CLSA) in which the boundary temperatures are assumed to be equal to the average temperatures. We use two-points Hermite approximations for integrals, based on the values of the integrand and its derivatives at the integration limits in the following form Mennig et al. (1983):

$$\int_a^b y(x)dx = \sum_{\nu=0}^{\alpha} C_\nu y^{(\nu)}(a) + \sum_{\nu=0}^{\beta} D_\nu y^{(\nu)}(b), \tag{16.14}$$

where $y(x)$ and its derivatives $y^{(\nu)}(x)$ are defined for all $x \in (a,b)$. It is assumed that the numerical values of $y^{(\nu)}(a)$ for $\nu = 0, 1, \ldots, \alpha$, and $y^{(\nu)}(b)$ for $\nu = 0, 1, \ldots, \beta$ are available. The general expression for the $H_{\alpha,\beta}$ approximation is given by

$$\begin{aligned}\int_a^b y(x)dx = &\sum_{\nu=0}^{\alpha} C_\nu(\alpha,\beta) h^{\nu+1} y^{(\nu)}(a) \\ &+ \sum_{\nu=0}^{\beta} C_\nu(\beta,\alpha)(-1)^\nu h^{\nu+1} y^{(\nu)}(b) + O\left(h^{\alpha+\beta+3}\right),\end{aligned} \tag{16.15}$$

where $h = b - a$, and

$$C_\nu(\alpha,\beta) = \frac{(\alpha+1)!(\alpha+\beta+1-\nu)!}{(\nu+1)!(\alpha-\nu)!(\alpha+\beta+2)!}. \tag{16.16}$$

We first employ the plain trapezoidal rule in the integrals for both average temperature and average heat flux ($H_{0,0}/H_{0,0}$ approximation), in the form

$$\begin{aligned}T_{av,i}(z,t) = &\frac{r_i}{r_{i+1}+r_i} T_i(r,z,t)|_{r=r_i} \\ &+ \frac{r_{i+1}}{r_{i+1}+r_i} T_i(r,z,t)|_{r=r_{i+1}}, \quad i = 1, 2, \ldots, N,\end{aligned} \tag{16.17}$$

$$T_i(r,z,t)|_{r=r_{i+1}} - T_i(r,z,t)|_{r=r_i} = \frac{r_{i+1}-r_i}{2}\left(\left.\frac{\partial T_i(r,z,t)}{\partial r}\right|_{r=r_i} + \left.\frac{\partial T_i(r,z,t)}{\partial r}\right|_{r=r_{i+1}}\right), \quad i=1,2,\ldots,N. \tag{16.18}$$

At each time t, analytical solution of the $4N$ unknowns $T_i(r,z,t)|_{r=r_i}$, $T_i(r,z,t)|_{r=r_{i+1}}$, $\partial T_i(r,z,t)/\partial r|_{r=r_i}$, and $\partial T_i(r,z,t)/\partial r|_{r=r_{i+1}}$, $i = 1,2,\ldots,N$, can be readily obtained by using a symbolic computation software such as *Mathematica* (Wolfram 2003) from a closed equation system formed by Eqs. (16.2)–(16.5), (16.17) and (16.18), and then used to close the partial differential equations Eq. (16.12) for the average temperature $T_{av,i}$, to be solved with the initial condition Eq. (16.13), providing the $H_{0,0}/H_{0,0}$ model.

Then we further improve the lumped model by employing the more accurate $H_{1,1}$ Hermite approximation (two-side corrected trapezoidal rule) in the integral for average temperature, in the form

$$T_{av,i}(z,t) = \left(\frac{1}{6}+\frac{r_i}{r_{i+1}+r_i}\right)T_i(r,z,t)|_{r=r_i} - \left(\frac{1}{6}-\frac{r_{i+1}}{r_{i+1}+r_i}\right)T_i(r,z,t)|_{r=r_{i+1}} + \frac{1}{6}\left(r_i\left.\frac{\partial T_i(r,z,t)}{\partial r}\right|_{r=r_i} - r_{i+1}\left.\frac{\partial T_i(r,z,t)}{\partial r}\right|_{r=r_{i+1}}\right), \quad i=1,2,\ldots,N, \tag{16.19}$$

while keeping the plain trapezoidal rule in the integral for heat flux ($H_{1,1}/H_{0,0}$ approximation). Similarly, the boundary temperatures and heat fluxes can be obtained from Eqs. (16.2)–(16.5), (16.18) and (16.19) and used to close the partial differential equations (16.12) for the average temperatures, to be solved with the initial condition Eq. (16.13), providing the $H_{1,1}/H_{0,0}$ model.

The improved lumped models governing the heat conduction in the composite pipeline and the transient energy equation (16.8) for the produced fluid are solved by using finite difference methods, for which the following discretization in the z-direction is introduced:

$$T_{av,i}(z,t) = T^j_{av,i}(t), \quad T_f(z,t) = T^j_f(t), \quad g_i(z,t) = g^j_i(t),$$
$$\frac{\partial T_f(z,t)}{\partial z} = \frac{T^j_f(t) - T^{j-1}_f(t)}{\Delta z}, \quad i=1,2,\ldots,N, \quad j=1,2,\ldots,Nz, \tag{16.20}$$

where Nz is the total number of grid points in the axial direction.

The discretized energy equation for the bulk temperature of produced fluid takes the following form:

$$\rho_f c_{pf} \frac{\mathrm{d}T_f^j(t)}{\mathrm{d}t} + \rho_f c_{pf} U_f \frac{T_f^j(t) - T_f^{j-1}(t)}{\Delta z} = \frac{h_1 P_w}{A_f}\left(T_1(r_1, z_j, t)|_{r=r_1} - T_f^j(t)\right),$$

$$\text{for} \quad t > 0, \qquad i = 1, 2, \ldots, N, \quad j = 1, 2, \ldots, Nz, \qquad (16.21)$$

The mathematical model consists of a system of $(N+1) \times N_z$ ordinary differential equations in time. At each discretized point z_j, there are N equations for the average temperatures in the multilayered composite pipeline, $T_{av,i}^j(t)$, and one equation for the bulk temperature $T_f^j(t)$. The ODE system can be readily solved numerically by using a well tested ODE solver. The Mathematica routine **NDSolve** was used for the solution of the ODE system.

16.4 Numerical Results and Discussions

As considered to be an effective solution for ultra-deepwater submarine multilayered composite pipelines combining high structural resistance with thermal insulation, sandwich pipes (SP) have attracted significant attention in recent years (Castello and Estefen 2007; An et al. 2012b, 2014). In this section, we employ the proposed improved lumped model to analyze the transient heat transfer in stainless steel–polypropylene–stainless steel SP ($N = 3$) with active electrical heating. According to Castello and Estefen (2007), Su et al. (2005), Bergman et al. (2011), Drescher et al. (2013), and Castello and Estefen (2008), the geometrical parameters and thermophysical properties of SP are presented in Table 16.1. Assume that the pipeline is with the length of 10 km and with the inlet boundary temperature of 90 °C, and the produced fluid is with the average velocity of 1.0 m/s and with the inlet boundary temperature of 90 °C. The sea water temperature T_m at the seabed is 4 °C. The linear heat generation rate of the innermost layer is 150 W/m.

Table 16.1 Geometrical parameters and thermophysical properties of sandwich pipes

r_1 (m)	0.07775[a]	c_{p2} (J/kg K)	2000[b]
r_2 (m)	0.08415[a]	ρ_3 (kg/m^3)	7850[b]
r_3 (m)	0.10315[a]	k_3 (W/m K)	54[b]
r_4 (m)	0.10955[a]	c_{p3} (J/kg K)	486[b]
ρ_1 (kg/m^3)	7850[b]	h_1 (W/m^2 K)	500[c]
k_1 (W/m K)	54[b]	h_2 (W/m^2 K)	150[d]
c_{p1} (J/kg K)	486[b]	ρ_f (kg/m^3)	800[e]
ρ_2 (kg/m^3)	775[b]	k_f (W/m K)	0.14[e]
k_2 (W/m K)	0.17[b]	c_{pf} (J/kg K)	2700[e]

[a], [b], [c], [d] and [e]: Refer to Castello and Estefen (2007), Su et al. (2005), Bergman et al. (2011), Drescher et al. (2013) and Castello and Estefen (2008), respectively

Table 16.2 Convergence behavior of the bulk temperature of the produced fluid T_f (°C) calculated by using $H_{0,0}/H_{0,0}$ improved lumped model

z (km)	$Nz = 100$	$Nz = 200$	$Nz = 300$	$Nz = 400$	$Nz = 500$
$t = 1$ h					
2	77.3255	78.1118	78.3083	78.3826	78.4176
4	31.9606	30.8308	30.4210	30.2148	30.0916
6	14.6757	14.4888	14.4307	14.4024	14.3856
8	12.2992	12.2739	12.2661	12.2623	12.2600
10	11.9776	11.9742	11.9731	11.9726	11.9723
$t = 2$ h					
2	78.3239	78.2935	78.2833	78.2782	78.2751
4	68.6123	68.9457	69.0014	69.0163	69.0213
6	43.5071	43.2243	43.0399	42.9132	42.8212
8	22.2046	21.6436	21.4733	21.3916	21.3437
10	18.4334	18.3555	18.3325	18.3214	18.3149
$t = 3$ h					
2	78.3239	78.2934	78.2833	78.2782	78.2751
4	69.0946	69.0466	69.0305	69.0224	69.0175
6	61.5947	61.7071	61.7126	61.7099	61.7068
8	47.9117	48.7405	49.1434	49.3857	49.5486
10	28.7392	27.7891	27.4498	27.2804	27.1798

Tables 16.2 and 16.3 present the convergence behavior of the bulk temperature of the produced fluid T_f calculated by using $H_{0,0}/H_{0,0}$, and $H_{1,1}/H_{0,0}$ improved lumped models against the number of grid points (Nz = 100, 200, 300, 400, and 500) along the length direction of pipelines, respectively. The bulk temperature at different positions (z = 2, 4, 6, 8, and 10 km) and different time (t = 1, 2, and 3 h) are listed. It can be observed that convergence is achieved essentially with a reasonably low number of grid points $Nz \leq 300$ for both $H_{0,0}/H_{0,0}$ and $H_{1,1}/H_{0,0}$ improved lumped models. Although convergence rates of the bulk temperature at t = 3 h are slower than the ones at t = 1 h, the absolute relative errors (ARE) are relatively low (i.e., 1.015% for the ARE of T_f at z = 10 km and t = 3 h calculated with Nz = 300 and Nz = 500 using $H_{1,1}/H_{0,0}$ improved lumped model). For the same cases, the bulk temperature profiles of the produced fluid at t = 1, 2 and 3 h calculated by using $H_{0,0}/H_{0,0}$ and $H_{1,1}/H_{0,0}$ improved lumped models are illustrated in Figs. 16.2 and 16.3.

In addition, the average temperature profile of each layer $T_{av,i}$ (°C) of SP at t = 3 h for i = 1, 2, and 3 calculated by using $H_{0,0}/H_{0,0}$, and $H_{1,1}/H_{0,0}$ improved lumped models are given in Figs. 16.4 and 16.5, from which it can be also seen that the good convergence is achieved. Figure 16.6 shows the bulk temperature distribution of the produced fluid T_f (°C) at different time (t = 0, 1, 2, 3, 4, 5, and 6) with various linear heat generation rate (0, 50, 100, and 150 W/m) calculated by using $H_{1,1}/H_{0,0}$ improved lumped model and the number of grid points Nz = 300. As the time increases, the temperature profile along the length direction approaches

Table 16.3 Convergence behavior of the bulk temperature of the produced fluid T_f (°C) calculated by using $H_{1,1}/H_{0,0}$ improved lumped model

z (km)	$Nz = 100$	$Nz = 200$	$Nz = 300$	$Nz = 400$	$Nz = 500$
$t = 1$ h					
2	77.0529	77.8289	78.0397	78.1292	78.1769
4	32.2292	31.1133	30.6988	30.4883	30.3621
6	14.7298	14.5374	14.4776	14.4485	14.4313
8	12.3156	12.2896	12.2815	12.2775	12.2752
10	11.9889	11.9854	11.9843	11.9837	11.9834
$t = 2$ h					
2	78.3244	78.2940	78.2837	78.2786	78.2755
4	68.3006	68.6867	68.7798	68.8180	68.8383
6	43.5050	43.3567	43.2613	43.1955	43.1474
8	22.3389	21.7570	21.5788	21.4933	21.4432
10	18.4421	18.3605	18.3363	18.3247	18.3179
$t = 3$ h					
2	78.3239	78.2934	78.2833	78.2782	78.2751
4	69.0938	69.0466	69.0305	69.0225	69.0176
6	61.3927	61.5675	61.6026	61.6154	61.6216
8	47.4949	48.2105	48.5496	48.7499	48.8828
10	28.9361	28.0196	27.6796	27.5060	27.4016

the steady-state temperature profile. The bulk temperature of the produced fluid at 10 km becomes 30.41 °C after 6 h, if no active heating is applied to SP. With active electrical heating, the bulk temperature at the same position rises up to 37.41, 44.40, and 51.40 °C after 6 h, when considering the linear heat generation rates of 50, 100, and 150 W/m, viz., the outlet temperature of the produced fluid increases by nearly 7, 14, and 21 °C, respectively.

Figure 16.7 shows the bulk temperature distribution of the produced fluid T_f (°C) at different time ($t = 0$, 1, 2, 3, 4, 5, and 6) with various average velocity ($U_f = 0.5$, 1.0, 1.5, and 2.0 m/s) calculated by using $H_{1,1}/H_{0,0}$ improved lumped model and the number of grid points $Nz = 300$. The linear heat generation rate of the innermost layer is 150 W/m. As the time increases, the temperature profile along the length direction approaches the steady-state temperature profile. Besides, the bulk temperature of the produced fluid reaches much more rapidly to its steady state as the average velocity increases. The bulk temperature at 10 km rises up to 31.44, 51.40, 59.63, and 65.14 °C after 6 h, respectively.

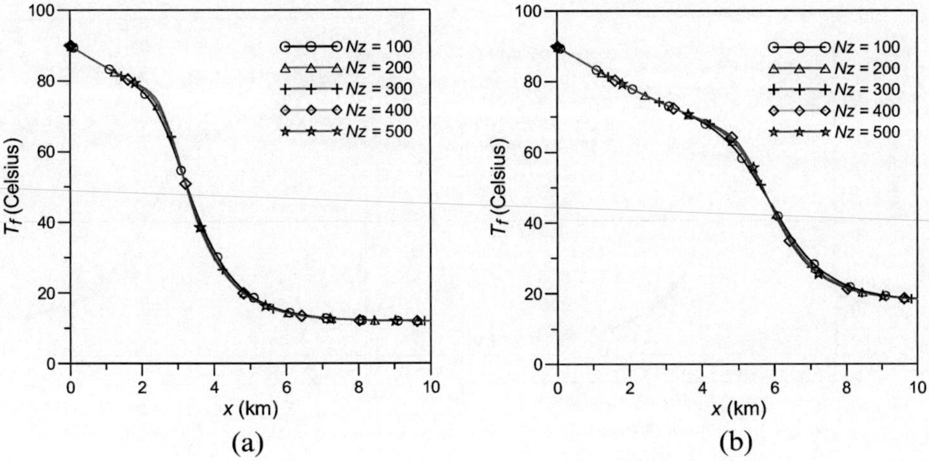

Fig. 16.2 Convergence behavior of the bulk temperature of the produced fluid T_f (°C) at $t = 1$, 2, and 3 h calculated by using $H_{0,0}/H_{0,0}$ improved lumped model. (**a**) $t = 1$ h. (**b**) $t = 2$ h. (**c**) $t = 3$ h

16.5 Conclusions

Based on two-point Hermite approximations for integrals, improved lumped parameter models were developed for the transient heat conduction of multilayered composite pipeline with active heating. The improved lumped models governing the heat conduction in the composite pipeline and the transient energy equation for the produced fluid were solved by using finite difference methods. Considering SP as an example, good convergence was achieved essentially with a reasonably low number of grid points $Nz \leq 300$ for both $H_{0,0}/H_{0,0}$ and $H_{1,1}/H_{0,0}$ improved lumped models. It was shown that by using active electrical heating with the linear heat

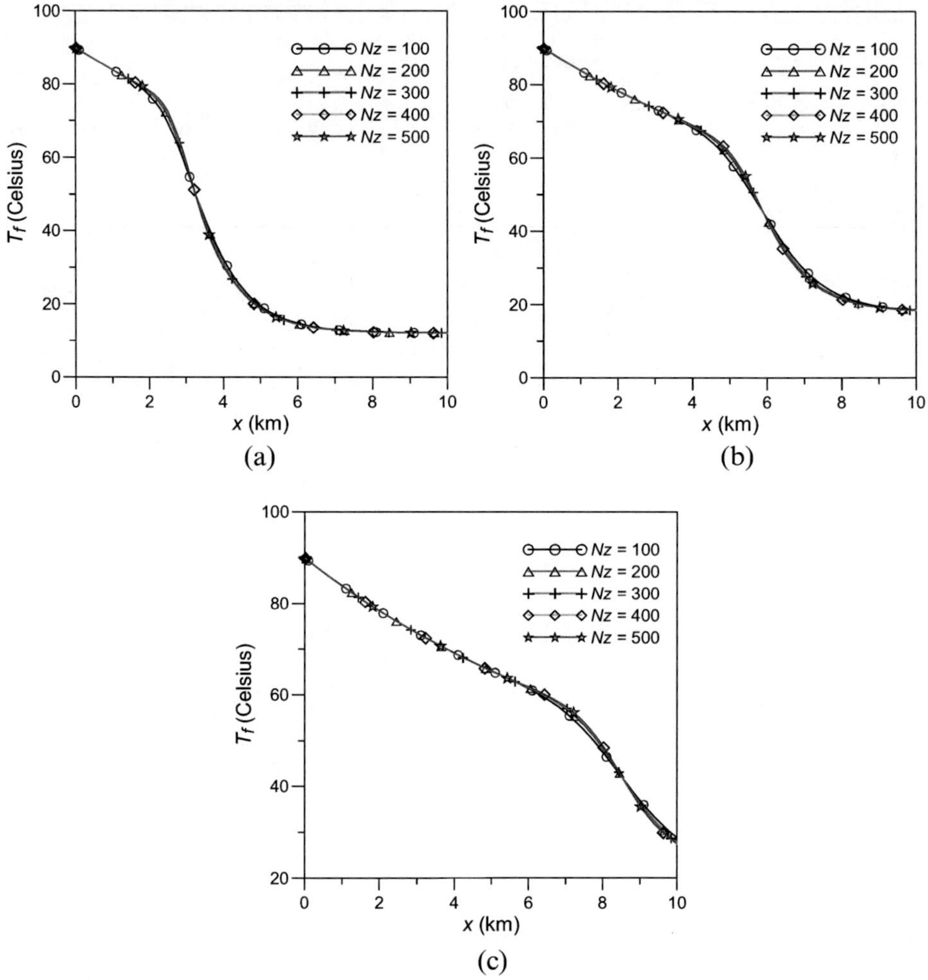

Fig. 16.3 Convergence behavior of the bulk temperature of the produced fluid T_f (°C) at $t = 1$, 2, and 3 h calculated by using $H_{1,1}/H_{0,0}$ improved lumped model. (**a**) $t = 1$ h. (**b**) $t = 2$ h. (**c**) $t = 3$ h

generation rates of 50, 100, and 150 W/m, the outlet temperature of the produced fluid increases by nearly 7, 14 and 21 °C, respectively. The bulk temperature of the produced fluid reached much more rapidly to its steady state as the average velocity increased. It was shown that the proposed improved lumped model approach can be utilized as an effective analytical tool for the thermal design and analysis of composite pipelines for oil and gas production in deepwater conditions.

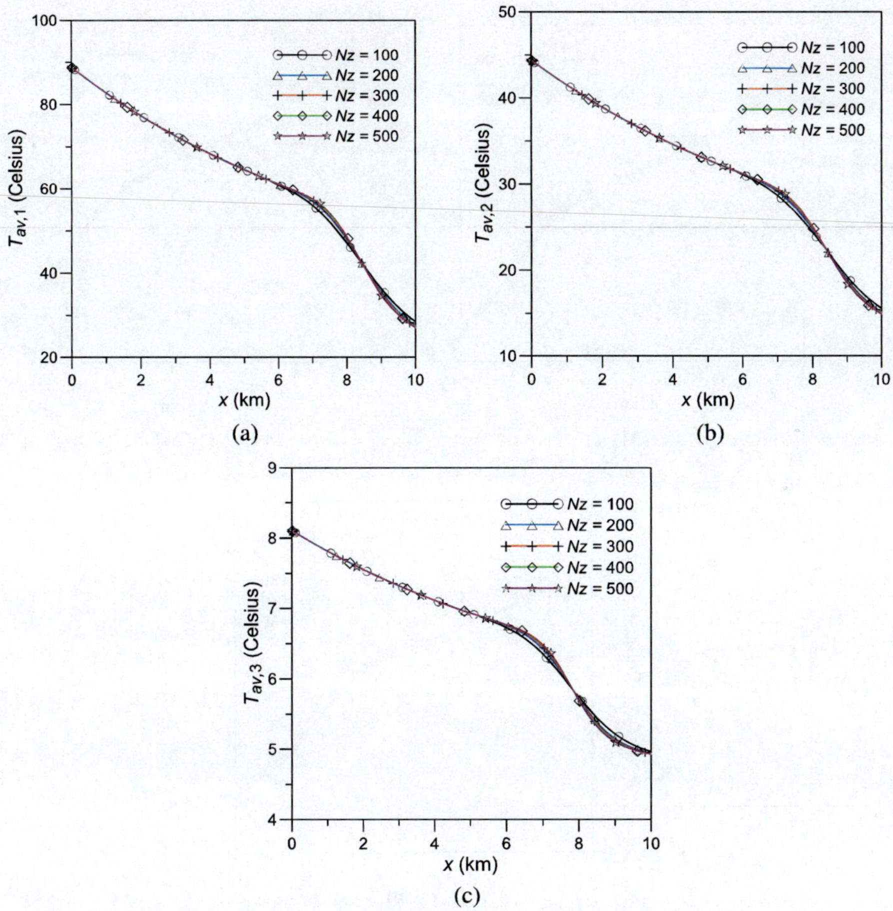

Fig. 16.4 Convergence behavior of the average temperature of each layer $T_{av,i}$ (°C) at $t = 3$ h for $i = 1$, 2, and 3 calculated by using $H_{0,0}/H_{0,0}$ improved lumped model. (**a**) $T_{av,1}$. (**b**) $T_{av,2}$. (**c**) $T_{av,3}$

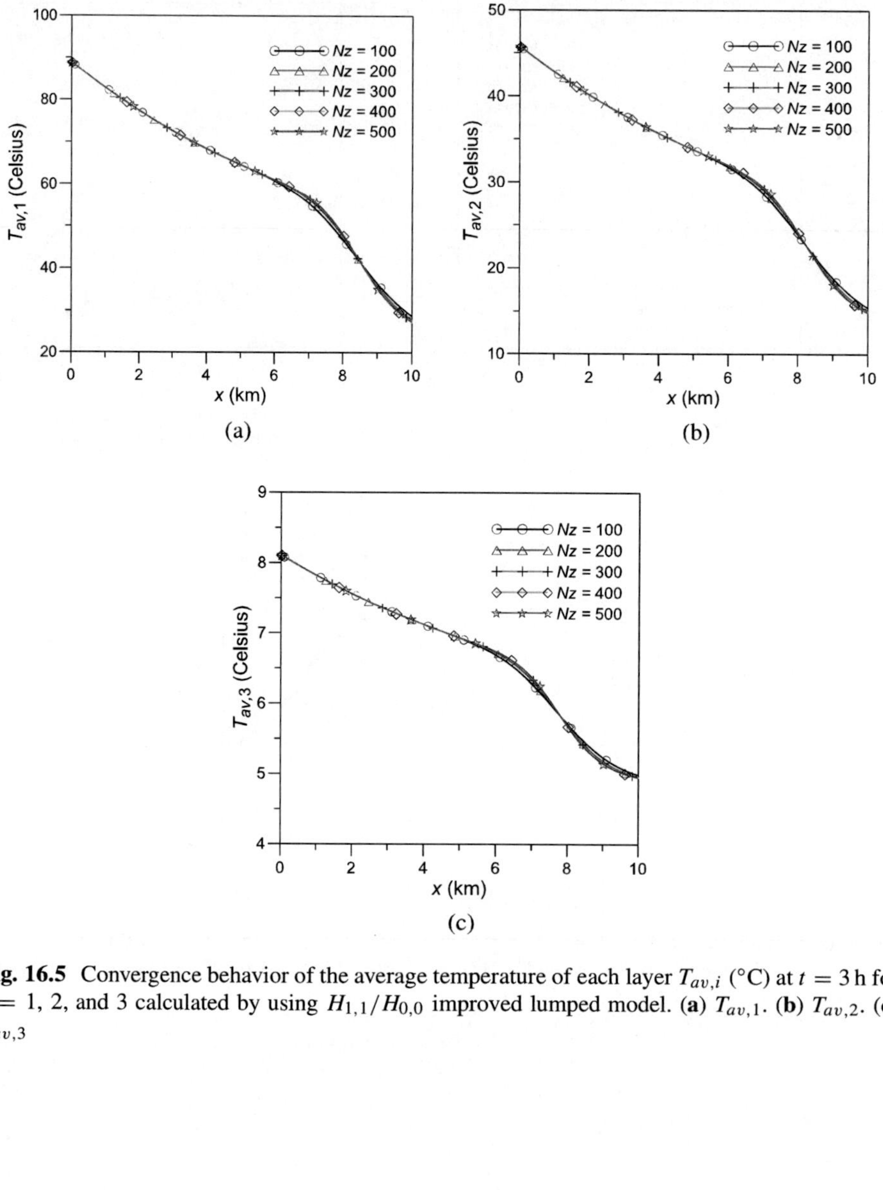

Fig. 16.5 Convergence behavior of the average temperature of each layer $T_{av,i}$ (°C) at $t = 3$ h for $i = 1$, 2, and 3 calculated by using $H_{1,1}/H_{0,0}$ improved lumped model. **(a)** $T_{av,1}$. **(b)** $T_{av,2}$. **(c)** $T_{av,3}$

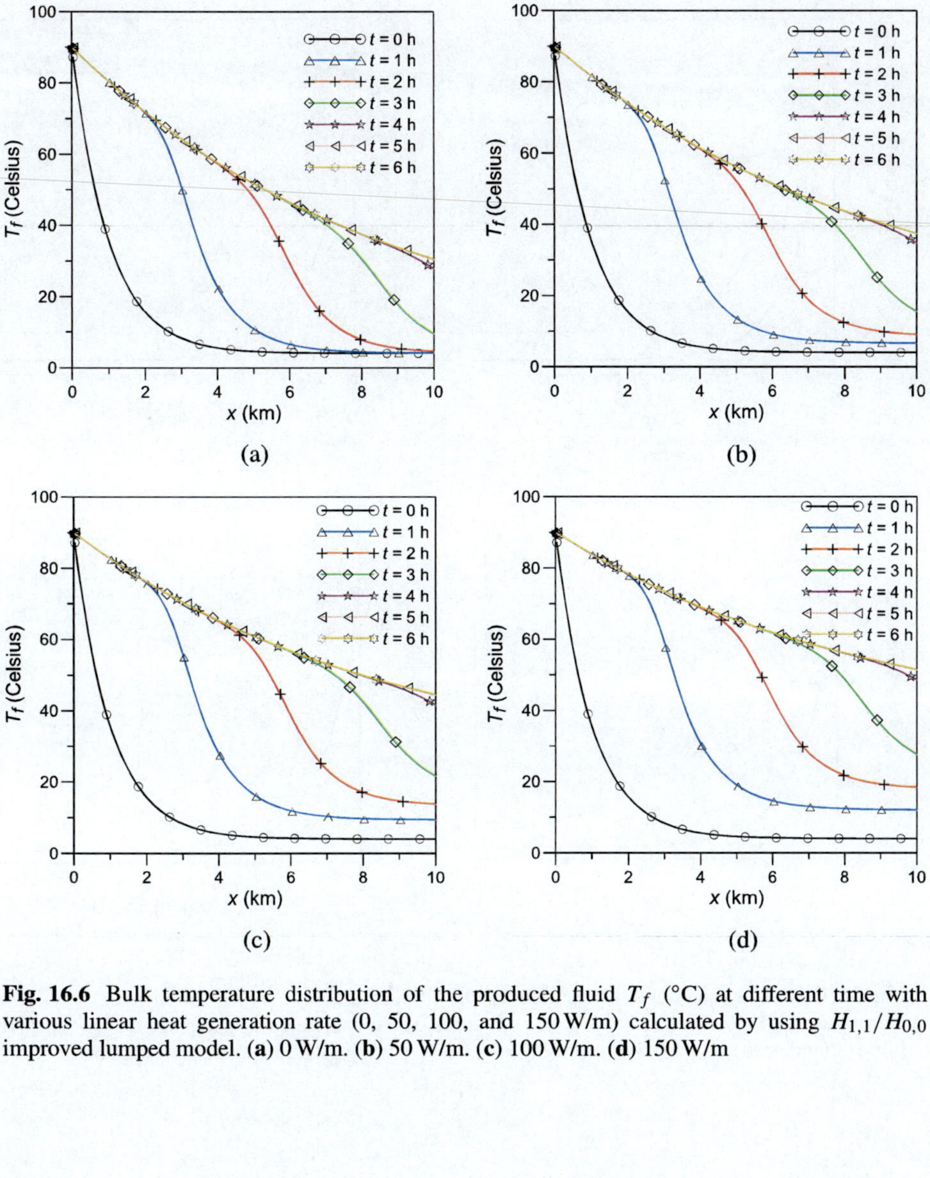

Fig. 16.6 Bulk temperature distribution of the produced fluid T_f (°C) at different time with various linear heat generation rate (0, 50, 100, and 150 W/m) calculated by using $H_{1,1}/H_{0,0}$ improved lumped model. (**a**) 0 W/m. (**b**) 50 W/m. (**c**) 100 W/m. (**d**) 150 W/m

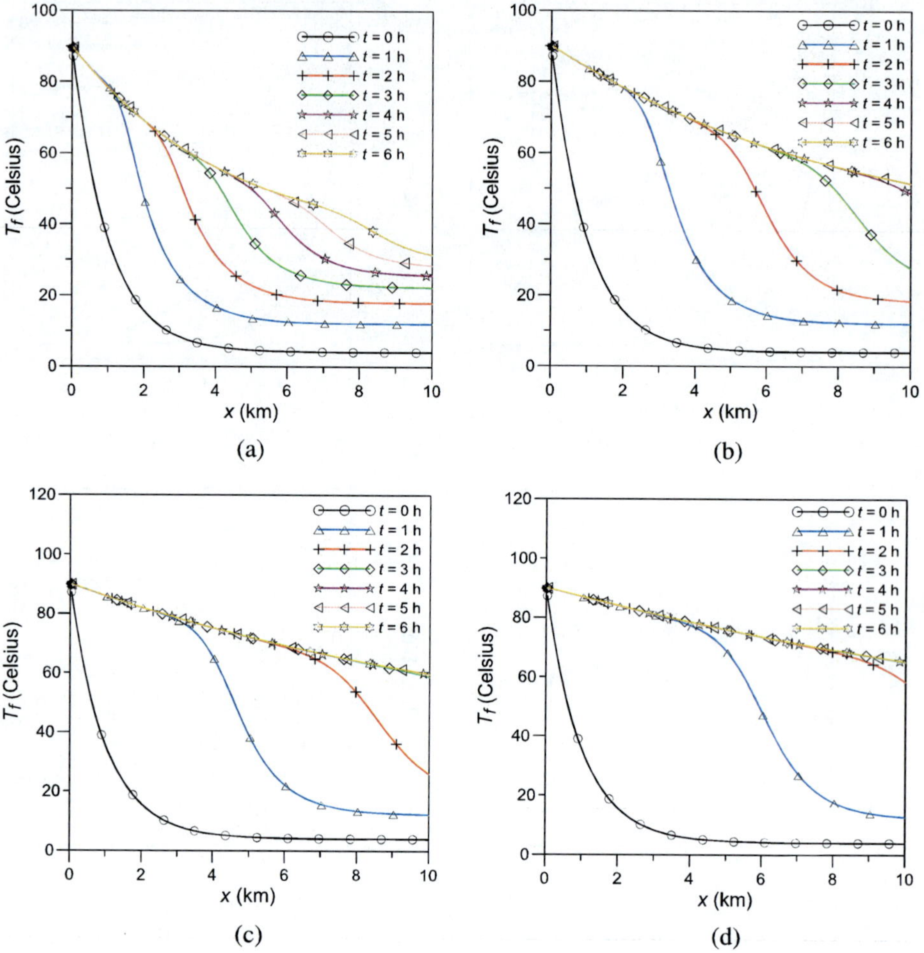

Fig. 16.7 Bulk temperature distribution of the produced fluid T_f (°C) at different time with various average velocity ($U_f = 0.5, 1.0, 1.5$, and 2.0 m/s) calculated by using $H_{1,1}/H_{0,0}$ improved lumped model. (**a**) $U_f = 0.5$ m/s. (**b**) $U_f = 1.0$ m/s. (**c**) $U_f = 1.5$ m/s. (**d**) $U_f = 2.0$ m/s

References

ABAQUS, *User's and Theory Manuals Version 6.9-1*. Hibbit, Karlsson and Sorensen, Inc, RI (2009a)

ABAQUS, *Example Problems Manual Version 6.9-1*. Hibbit, Karlsson and Sorensen, Inc, RI (2009b)

ABAQUS, *User's and Theory Manuals*. Hibitt, Karlsson, Sorensen (2013)

F. Albermani, H. Khalilpasha, H. Karampour, Propagation buckling in deep sub-sea pipelines. Eng. Struct. **33**(9), 2547–2553, (2011)

A.E. Alshorbagy, M.A. Eltaher, F.F. Mahmoud, Free vibration characteristics of a functionally graded beam by finite element method. Appl. Math. Model. **35**(1), 412–425 (2011)

C. An, Colapso de Dutos Sanduíche com Núcleo de Compósitos Cimentícios Reforçados com Fibras de PVA Sob Pressão Hidrostática. PhD Thesis, COPPE/UFRJ. Rio de Janeiro, 2012

C. An, J. Su, Improved lumped models for transient combined convective and radiative cooling of multi-layer composite slabs. Appl. Therm. Eng. **31**(14–15), 2508–2517 (2011)

C. An, J. Su, Lumped parameter model for one-dimensional melting in a slab with volumetric heat generation. Appl. Therm. Eng. **60**(1–2), 387–396 (2013)

C. An, J. Su. Dynamic analysis of axially moving orthotropic plates: integral transform solution. Appl. Math. Comput. **228**, 489–507 (2014a)

C. An, J. Su, Dynamic response of axially moving Timoshenko beams: integral transform solution. Appl. Math. Mech. **35**, 1–16 (2014b)

C. An, J. Su, Dynamic behavior of pipes conveying gas-liquid two-phase flow. Nuclear Eng. Design **292**, 204–212 (2015)

C. An, J. Su, Dynamic behavior of axially functionally graded pipes conveying fluid. Math. Probl. Eng. **2017**, 6789634 (2017)

C. An, X. Castello, M.O. Andrielli, M.L. Duan, R.D. Toledo Filho, S.F. Estefen, Limit strength of new sandwich pipes with strain hardening cementitious composites (SHCC) core: Finite element modelling, in *Proceedings of the 31st International Conference on Ocean, Offshore and Arctic Engineering*, Rio de Janeiro (2012a)

C. An, X. Castello, M.L. Duan, R.D. Toledo, S.F. Estefen, Ultimate strength behaviour of sandwich pipes filled with steel fiber reinforced concrete. Ocean Eng. **55**, 125–135 (2012b)

C. An, X. Castello, M.L. Duan, R.D. Toledo, S.F. Estefen, Ultimate strength behaviour of sandwich pipes filled with steel fiber reinforced concrete. Ocean Eng. **55**, 125–135 (2012c)

C. An, M.L. Duan, S.F. Estefen. Collapse and buckle propagation of sandwich pipes: A review, in *Proceedings of the 32nd International Conference on Ocean, Offshore and Arctic Engineering*, Nantes (2013)

C. An et al., *Structural and Thermal Analyses of Deepwater Pipes*,
https://doi.org/10.1007/978-3-030-53540-7

C. An, M.L. Duan, R.D. Toledo, S.F. Estefen, Collapse of sandwich pipes with PVA fiber reinforced cementitious composites core under external pressure. Ocean Eng. **82**, 1–13 (2014)

C. An, J. Gu, J. Su. Exact solution of bending problem of clamped orthotropic rectangular thin plates. J. Braz. Soc. Mech. Sci. Eng. **38**, 1–7 (2015)

G. Anderson. On determination of finite integral transforms for forced vibrations of circular plates. J. Sound Vib. **9**(1), 126–144 (1969)

K. Arjomandi, F. Taheri, Elastic buckling capacity of bonded and unbonded sandwich pipes under external hydrostatic pressure. J. Mech. Mater. Struct. **5**(3), 391–408 (2010)

K. Arjomandi, F. Taheri, The influence of intra-layer adhesion configuration on the pressure capacity and optimized configuration of sandwich pipes. Ocean Eng. **38**(17–18), 1869–1882 (2011a)

K. Arjomandi, F. Taheri, A new look at the external pressure capacity of sandwich pipes. Marine Struct. **24**(1), 23–42 (2011b)

K. Arjomandi, F. Taheri, Stability and post-buckling response of sandwich pipes under hydrostatic external pressure. Int. J. Pres. Ves. Pip. **88**(4), 138–148 (2011c)

K. Arjomandi, F. Taheri, Bending capacity of sandwich pipes. Ocean Eng. **48**, 17–31 (2012)

Y. Bai, Q. Bai, *Subsea Pipeline and Risers* (Elsevier, Oxford, 2005)

P.N. Balaguru, S.P. Shah, *Fiber-Reinforced Cement Composites* (McGraw-Hill, New York, 1992)

D. Barnea, Effect of bubble shape on pressure drop calculations in vertical slug flow. Int. J. Multip. Flow **16**(1), 79–89 (1990)

K.E. Barth, H.Y. Wu, Efficient nonlinear finite element modeling of slab on steel stringer bridges. Finite Elem. Analy. Des. **42**, 1304–1313 (2006)

Y. Basogul, A. Kecebas, Economic and environmental impacts of insulation in district heating pipelines. Energy **36**(10), 6156–6164 (2011)

T.L. Bergman, A.S. Lavine, F.P. Incropera, D.P. DeWitt, *Fundamentals of Heat and Mass Transfer*, 7th edn. (Wiley, New York, 2011)

A. Blanco, E. Casanova, A preliminary study on the interaction between slug flow and vortex induced vibration in fatigue life of submarines pipelines, in *Proceedings of the ASME 29th International Conference on Ocean, Offshore and Arctic Engineering* (2010)

W.P. Boshoff, G.P.A.G. Van Zijl, A computational model for strain-hardening fibre-reinforced cement-based composites. J. S. Afr. Inst. Civ. Eng. **49**(2), 24–31 (2007)

B.M. Bossio, A.J. Blanco, E.L. Casanova, Numerical modeling of the dynamical interaction between slug flow and vortex induced vibration in horizontal submarine pipelines. J. Offshore Mech. Arct. Eng. Trans. **136**(4), 041803 (2014)

M.B. Bryndum, P.F. Smed, Application of the generalized force model for VIV in pipeline span design, in *Proceedings of the ASME 17th International Conference on Ocean, Offshore and Arctic Engineering* (1998)

E.L. Casanova, A.J. Blanco, Effects of soil non-linearity on the dynamic behavior and fatigue life of pipeline spans subjected to slug flow, in *Proceedings of the ASME 29th International Conference on Ocean, Offshore and Arctic Engineering* (2010)

X. Castello, Efeitos do Processo de Instalação na Resistência Estrutural de Dutos Sanduíche. PhD Thesis, COPPE/UFRJ. Rio de Janeiro, 2005

X. Castello, Influência da Adesão Entre Camadas na Resistência ao Colapso de Dutos Sanduíche. PhD Thesis, COPPE/UFRJ. Rio de Janeiro (2011a)

X. Castello, Adehsion between layers and coupling of sandwich pipes. PhD thesis, COPPE, Federal University of Rio de Janeiro, Rio de Janeiro (2011b)

X. Castello, S.F. Estefen, Limit strength and reeling effects of sandwich pipes with bonded layers. Int. J. Mech. Sci. **49**(5), 577–588 (2007)

X. Castello, S.F. Estefen, Sandwich pipes for ultra deepwater applications, in *Proceedings of Offshore Technology Conference*, Houston, Texas (2008)

X. Castello, S.F. Estefen, H.R. Leon, L.C. Chad, J. Souza, Design aspects and benefits of sandwich pipes for ultra deepwaters, in *Proceedings of the 28th International Conference on Ocean, Offshore and Arctic Engineering*, Honolulu, HI (2009)

D.R. Cerqueira, J. Su, S.F. Estefen, Simulation of transient heat transfer of sandwich pipes with active electrical heating, in *Proceedings of 23rd International Conference on Offshore Mechanics and Arctic Engineering (OMAE'04)*, Vancouver (2004)

Y.S. Cha, W.J. Minkowycz, S.Y. Seol, Transverse temperature distribution and heat-generation rate in composite superconductors subjected to constant thermal disturbance. Int. Commun. Heat Mass **22**(4), 461–474 (1995)

I.K. Chatjigeorgiou, On the effect of internal flow on vibrating catenary risers in three dimensions. Eng. Struct. **32**(10), 3313–3329 (2010a)

I.K. Chatjigeorgiou, Three dimensional nonlinear dynamics of submerged, extensible catenary pipes conveying fluid and subjected to end-imposed excitations. Int. J. Non-Linear Mech. **45**(7), 667–680 (2010b)

Y.M. Chen, S.W. Wang, Z. Zuo, A procedure for calculating transient thermal load through multilayer cylindrical structures. Appl. Therm. Eng. **23**(16), 2133–2145 (2003)

L.Q. Chen, Y.L. Zhang, G.C. Zhang, H. Ding, Evolution of the double-jumping in pipes conveying fluid flowing at the supercritical speed. In. J. Non-Linear Mech. **58**, 11–21 (2014)

D. Chisholm, *Two-Phase Flow in Pipelines and Heat Exchangers* (George Godwin, Boca Raton, 1983)

R.M. Cotta, *Integral Transforms in Computational Heat and Fluid Flow* (CRC Press, Boca Raton, 1993)

R.M. Cotta, *The Integral Transform Method in Thermal and Fluids Science and Engineering* (Begell House, New York, 1998)

R.M. Cotta, M.D. Mikhailov, *Heat Conduction - Lumped Analysis, Integral Transforms, Symbolic Computation* (Wiley, Chichester, 1997)

H.L. Dai, L. Wang, Q. Qian, Q. Ni, Vortex-induced vibrations of pipes conveying fluid in the subcritical and supercritical regimes. J. Fluids Struct. **39**, 322–334 (2013)

H.L. Dai, A. Abdelkefi, L. Wang, Modeling and nonlinear dynamics of fluid-conveying risers under hybrid excitations. Int. J. Eng. Sci. **81**, 1–14 (2014a)

H.L. Dai, L. Wang, Q. Qian, Q. Ni, Vortex-induced vibrations of pipes conveying pulsating fluid. Ocean Eng. **77**, 12–22 (2014b)

R. de Borst, H.-B. Mühlhaus, Gradient-dependent plasticity: formulation and algorithmic aspects. Int. J. Num. Method Eng. **35**(3), 521–539 (1992)

F. de Monte, An analytic approach to the unsteady heat conduction processes in one-dimensional composite media. Int. J. Heat Mass Tran. **45**(6), 1333–1343 (2002)

A.R. de Souza, T.A. Netto, I.P. Pasqualino, Materials selection for sandwich pipes under the combined effect of pressure, bending and temperature, in *Proceedings of 26th International Conference on Offshore Mechanics and Arctic Engineering (OMAE'07)*, San Diego (2007)

D. Deka, M. Cerkovnik, N. Panicker, V. Achanta. Subsea jumpers vibration assessment, in *Proceedings of the ASME 32nd International Conference on Ocean, Offshore and Arctic Engineering* (2013)

A.A. Delouei, M.H. Kayhani, M. Norouzi, Exact analytical solution of unsteady axi-symmetric conductive heat transfer in cylindrical orthotropic composite laminates. Int. J. Heat Mass Trans. **55**(15–16), 4427–4436 (2012)

Diana, *Displacement Analysis Finite Element Software Version 9.2*. TNO-Building Division, Delf, (2000)

M. Drescher, B. Wilhelmsen, P. Aursand, E. Aursand, G. de Koeijer, R. Held. Heat transfer characteristics of a pipeline for CO_2 transport with water as surrounding substance. Energy Procedia **37**, 3047–3056 (2013)

J.Y. Dyau, S. Kyriakides, On the propagation pressure of long cylindrical shells under external pressure. Int. J. Mech. Sci. **35**(8), 675–713 (1993)

M. Elchalakani, X.L. Zhao, R. Grzebieta, Tests on concrete filled double-skin (CHS outer and SHS inner) composite short columns under axial compression. Thin-Walled Struct. **40**(5), 415–441 (2002)

S.F. Estefen, Collapse behaviour of intact and damaged deepwater pipelines and the influence of the reeling method of installation. J. Construct. Steel Res. **50**(2), 99–114 (1999)

S.F. Estefen, T. Moan, S. Saevik, R.A. Zimmer, Limit state formulations for TLP Tendon and Steel Riser bodies. J. Construct. Steel Res. **32**(1), 107–121 (1995)

S.F. Estefen, T.A. Netto, I.P. Pasqualino. Strength analyses of sandwich pipes for ultra deepwaters. J. Appl. Mech. Trans. **72**(4), 599–608 (2005a)

M.L. Facchinetti, E. de Langre, F. Biolley, Coupling of structure and wake oscillators in vortex-induced vibrations. J. Fluids Struct. **19**(2), 123–140 (2004)

R.C. Fernandes, R. Semiat, A.E. Dukler, Hydrodynamic model for gas-liquid slug flow in vertical tubes. AIChE J. **29**(6), 981–989 (1983)

J.F. Franklin, A. Wright, The development of thermal insulation systems for deep water pipelines, in *Proceedings 13th International Conference on Pipeline Protection*, Edinburgh (1999), pp. 217–229

G. Fu, C. Paz, J. Chujutalli, M.I. Lourenço, D. de Lima Jr, Y. Li, R. Toledo Filho, S. Estefen, Sandwich pipes with strain hardening cementitious composites (SHCC): Numerical Analysis, in *International Conference on Offshore Mechanics and Arctic Engineering* (2014)

G.K. Furnes, J. Berntsen, On the response of a free span pipeline subjected to ocean currents. Ocean Eng. **30**(12), 1553–1577 (2003)

N. Ganesan, R. Kadoli, A study on the dynamic stability of a cylindrical shell conveying a pulsatile flow of hot fluid. J. Sound Vib. **274**(3–5), 953–984 (2004)

F. Garcia, R. Garcia, J.C. Padrino, C. Mata, J.L. Trallero, D.D. Joseph, Power law and composite power law friction factor correlations for laminar and turbulent gas-liquid flow in horizontal pipelines. Int. J. Multiphase Flow **29**, 1605–1624 (2003)

J.H. Ginsberg, The dynamic stability of a pipe conveying a pulsatile flow. Int. J. Eng. Sci. **11**, 1013–1024 (1973)

S. Gong, B. Sun, S. Bao, Y. Bai, Buckle propagation of offshore pipelines under external pressure. Marine Struct. **29**(1), 115–130 (2012)

S.-F. Gong, X.-Y. Ni, S. Bao, Y. Bai, Asymmetric collapse of offshore pipelines under external pressure. Ships Offshore Struct. **8**(2), 176–188 (2013)

C.D. Goode, P. Montague, T. Nash. Large reinforced concrete cylinders under external pressure. Proc. Inst. Civil Eng. Struct. Build. **116**(2), 163–173 (1996)

J.J. Gu, C. An, C. Levi, J. Su, Prediction of vortex-induced vibration of long flexible cylinders modeled by a coupled nonlinear oscillator: integral transform solution. J. Hydrodyn. **24**(6), 888–898 (2012)

J.J. Gu, C. An, M.L. Duan, C. Levi, J. Su, Integral transform solutions of dynamic response of a clamped-clamped pipe conveying fluid. Nuclear Eng. Design **254**, 237–245 (2013)

V.I. Gulyayev, E.Y. Tolbatov, Forced and self-excited vibrations of pipes containing mobile boiling fluid clots. J. Sound Vib. **257**(3), 425–437 (2002)

H.Y. Guo, S.Q. Wang, J.N. Wu, D.F. Liu, Dynamic characteristics of marine risers conveying fluid. China Ocean Eng. **14**(2), 153–160 (2000)

H.Y. Guo, Y.B. Wang, Q. Fu, The effect of internal fluid on the response of vortex-induced vibration of marine risers. China Ocean Eng. **18**(1), 11–20 (2004)

H.Y. Guo, X.M. Li, X.C. Liu, Numerical prediction of vortex-induced vibrations on top tensioned riser in consideration of internal flow. China Ocean Eng. **22**(4), 675–682 (2008)

B.Y. Guo, S.H. Song, A. Ghalambor, T.R. Lin, *Offshore Pipelines, Second Edition: Design, Installation and Maintenance* (Elsevier, Oxford, 2014)

A. Haji-Sheikh, J.V. Beck, Greens-function partitioning in Galerkin-based integral solution of the diffusion equation. J. Heat Trans. **112**(1), 28–34 (1990)

V.H. Halvorsen, J.K. Lervik, G. Klevjer, Hydrate and was prevention of risers by electrical methods, in *The Tenth International Offshore and Polar Engineering Conference, 28 May-2 June, Seattle, Washington* (International Society of Offshore and Polar Engineers, Mountain View, 2000), pp. 42–48

J.H. Han, G.A. Kardomateas, G.J. Simitses, Elasticity, shell theory and finite element results for the buckling of long sandwich cylindrical shells under external pressure. Compos. Part B-Eng. **35**(6–8), 591–598 (2004a)

L.H. Han, Z. Tao, H. Huang, X.L. Zhao, Concrete-filled double skin (SHS outer and CHS inner) steel tubular beam-columns. Thin-Walled Struct. **42**(9), 1329–1355 (2004b)

L.H. Han, G.H. Yao, Z. Tao, Performance of concrete-filled thin-walled steel tubes under pure torsion. Thin-Walled Struct. **45**(1), 24–36 (2007)

L.H. Han, H. Huang, X.L. Zhao, Analytical behaviour of concrete-filled double skin steel tubular (CFDST) beam-columns under cyclic loading. Thin-Walled Struct. **47**(6–7), 668–680 (2009)

F. Hara, Two-phase-flow-induced vibrations in a horizontal piping system. Bull. Jpn Soc. Mech. Eng. **20**(142), 419–427 (1977)

G.E. Harrison, D.G. Herring, Active heating, dual production flowlines design for king sub sea tie-back at 1670 meters water depth, in *Proceedings of ETCE/OMAE 2000 Joint Conference, Paper No. OMAE* (2000)

T. He, M. Duan, J. Wang, S. Lv, C. An, On the external pressure capacity of deepwater sandwich pipes with inter-layer adhesion conditions. Appl. Ocean Res. **52**, 115–124 (2015)

K. Holschemacher, T. Mueller, Y. Ribakov, Effect of steel fibres on mechanical properties of high-strength concrete. Mater. Des. **31**(5), 2604–2615 (2010)

M. Hosseini, S.A. Fazelzadeh, Thermomechanical stability analysis of functionally graded thin-walled cantilever pipe with flowing fluid subjected to axial load. Int. J. Struct. Stabil. Dyn. **11**(3), 513–534 (2011)

Y. Huang, X.F. Li, A new approach for free vibration of axially functionally graded beams with non-uniform cross-section. J. Sound Vib. **329**(11), 2291–2303 (2010)

H. Huang, L.H. Han, Z. Tao, X.L. Zhao, Analytical behaviour of concrete-filled double skin steel tubular (CFDST) stub columns. J. Construct. Steel Res. **66**(4), 542–555 (2010a)

Y.M. Huang, Y.S. Liu, B.H. Li, Y.J. Li, Z.F. Yue, Natural frequency analysis of fluid conveying pipeline with different boundary conditions. Nuclear Eng. Design **240**(3), 461–467 (2010b)

Y. Huang, L.E. Yang, Q.Z. Luo, Free vibration of axially functionally graded Timoshenko beams with non-uniform cross-section. Compos. Part B-Eng. **45**(1), 1493–1498 (2013)

IMSL, *IMSL Fortran Library Version 5.0, Math/Library*, vol. 2 (Visual Numerics, Houston, 2003)

P.K. Jain, S. Singh, R. Uddin, An exact analytical solution for two-dimensional, unsteady, multilayer heat conduction in spherical coordinates. Int. J. Heat Mass Tran. **53**(9–10), 2133–2142 (2010)

J.D. Jin, Z.Y. Song, Parametric resonances of supported pipes conveying pulsating fluid. J. Fluids Struct. **20**(6), 763–783 (2005)

P. Jun, V. Mechtcherine, Behaviour of strain-hardening cement-based composites (SHCC) under monotonic and cyclic tensile loading part 1-experimental investigations. Cem. Concr. Compos. **32**(10), 801–809 (2010)

R. Kadoli, N. Ganesan, Parametric resonance of a composite cylindrical shell containing pulsatile flow of hot fluid. Compos. Struct. **65**(3–4), 391–404 (2004)

S. Kaewunruen, J. Chiravatchradej, S. Chucheepsakul, Nonlinear free vibrations of marine risers/pipes transporting fluid. Ocean Eng. **32**(3–4), 417–440 (2005)

M.-G. Kang, The influence of rotary inertia of concentrated masses on the natural vibrations of a clamped-supported pipe conveying fluid. Nuclear Eng. Design **196**(3), 281–292 (2000)

G.A. Kardomateas, Elasticity solutions for sandwich orthotropic cylindrical shells under external/internal pressure or axial force. AIAA J. **39**(4), 713–719 (2001)

G.A. Kardomateas, G.J. Simitses, Buckling of long sandwich cylindrical shells under external pressure. J. Appl. Mech. Trans. **72**(4), 493–499 (2005)

M.H. Kayhani, M. Norouzi, A. Amiri Delouei, A general analytical solution for heat conduction in cylindrical multilayer composite laminates. Int. J.Thermal Sci. **52**(1), 73–82 (2012)

H. Khalilpasha, F. Albermani, Hyperbaric chamber test of subsea pipelines. Thin-Walled Struct. **71**, 1–6 (2013)

T.S. Kim, H. Kuwamura, Finite element modeling of bolted connections in thin-walled stainless steel plates under static shear. Thin-Walled Struct. **45**(4), 407–421 (2007)

G.L. Kuiper, A.V. Metrikine, Dynamic stability of a submerged, free-hanging riser conveying fluid. J. Sound Vib. **280**(3–5), 1051–1065 (2005)

S. Kyriakides, Buckle propagation in pipe-in-pipe systems.: Part I. Experiments. Int. J. Solids Struct. **39**(2), 351–366 (2002)

S. Kyriakides, C.D. Babcock, Experimental determination of the propagation pressure of circular pipes. J. Press. Vessel Technol. **103**(4), 328–336 (1981)

S. Kyriakides, E. Corona, *Mechanics of Offshore Pipelines. Volume 1: Buckling and Collapse* (Elsevier, Oxford, 2007a)

S. Kyriakides, T.A. Netto, On the dynamic propagation and arrest of buckles in pipe-in-pipe systems. Int. J. Solids Struct. **41**(20), 5463–5482 (2004)

S. Kyriakides, T.J. Vogler, Buckle propagation in pipe-in-pipe systems. Part II. Analysis. Int. J. Solids Struct. **39**(2), 367–392 (2002)

N. Laoutr, S. Denntel, Heated PIP: combining passive insulation and active heating for enhanced thermal performance, in *Proceedings of the 20th International Conference on Offshore Mechanics and Arctic Engineering–OMAE* (2001)

A.W. Leissa, M.S. Qatu, *Vibrations of Continuous Systems* (McGraw-Hill, New York, 2011)

J.K. Lervik, H. Kulbotten, G. Klevjer, Prevention of hydrate formation in pipelines by electrical methods, in *International Offshore and Polar Engineering Conference Proceedings* (1997), pp. 249–254

J.K. Lervik, M. Ahlbeck, H. Raphael, T. Lauvdal, P. Holen, Direct electrical heating of pipelines as a method of preventing hydrate and wax plugs, in *International Offshore and Polar Engineering Conference Proceedings* (1998), pp. 39–45

V.C. Li, From micromechanics to structural engineering - the design of cementitious composites for civil engineering applications, in *Proceedings of the Japan Society of Civil Engineers* **10**(2), 37–48 (1994)

Q.H. Li, S.L. Xu, Experimental research on mechanical performance of hybrid fiber reinforced cementitious composites with polyvinyl alcohol short fiber and carbon textile. J. Compos. Mater. **45**(1), 5–28 (2010)

M. Li, A.C.K. Lai, Analytical solution to heat conduction in finite hollow composite cylinders with a general boundary condition. Int. J. Heat Mass Trans. **60**(1), 549–556, (2013)

V.C. Li, C.K.Y. Leung, Steady-state and multiple cracking of short random fiber composites. J. Eng. Mech. **118**(11), 2246–2264 (1992)

V.C. Li, S.X. Wang, C. Wu, Tensile strain-hardening behavior of polyvinyl alcohol engineered cementitious composite (PVA-ECC). ACI Mater. J. **98**(6), 483–492 (2001)

B.H. Li, H.S. Gao, H.B. Zhai, Y.S. Liu, Z.F. Yue, Free vibration analysis of multi-span pipe conveying fluid with dynamic stiffness method. Nuclear Eng. Design **241**(3), 666–671 (2011)

F. Liang, B.C. Wen, Forced vibrations with internal resonance of a pipe conveying fluid under external periodic excitation. Acta Mech. Solida Sin. **24**(6), 477–483 (2011)

P. Lima, Análise Teórica e experimental de Compósitos Reforçados com Fibras de Sisal. PhD Thesis, COPPE/UFRJ. Rio de Janeiro, 2004

L. Liu, F. Xuan, Flow-induced vibration analysis of supported pipes conveying pulsating fluid using precise integration method. Math. Probl. Eng. **2010**, 806475 (2010)

R.H. Long, Experimental and theoretical study of transverse vibration of a tube containing flowing fluid. J. Appl. Mechan. Trans. **22**(1), 65–68 (1955)

M.I. Lourenço, I.P. Pasqualino, T.S. Palermo, Core Material Performance on the Propagation Pressure of Sandwich Pipes, in Proceedings of the 27th International Conference on Ocean, Offshore and Arctic Engineering, Estoril, Portugal (2008)

X.B. Lu, C.T.T. Hsu, Behavior of high strength concrete with and without steel fiber reinforcement in triaxial compression. Cement Concrete Res. **36**(9), 1679–1685 (2006)

I.R. Maestre, P.R. Cubillas, L. Perez-Lombard, Transient heat conduction in multi-layer walls: an efficient strategy for Laplace's method. Energy Build. **42**(4), 541–546 (2010)

C.F.T. Matt, On the application of generalized integral transform technique to wind-induced vibrations on overhead conductors. Int. J. Numer. Method Eng. **78**(8), 901–930 (2009)

C.F.T. Matt, Combined classical and generalized integral transform approaches for the analysis of the dynamic behavior of a damaged structure. Appl. Math. Model. **37**(18–19), 8431–8450 (2013a)

C.F.T. Matt, Simulation of the transverse vibrations of a cantilever beam with an eccentric tip mass in the axial direction using integral transforms. Appl. Math. Model. **37**(22), 9338–9354 (2013b)

J. Mennig, T. Auerbach, W. Hälg, Two point Hermite approximation for the solution of linear initial value and boundary value problems. Comp. Meth. Appl. Mech. Eng. **39**, 199–224 (1983)

M.D. Mikhailov, M.N. Özişik, *Unified Analysis and Solutions of Heat and Mass Diffusion* (Wiley, New York, 1984)

M.D. Mikhailov, M.N. Özişik, N.L. Vulchanov, Diffusion in composite layers with automatic solution of the eigenvalue problem. Int. J. Heat Mass Tran. **26**(8), 1131–1141 (1983)

C. Monette, M.J. Pettigrew, Fluidelastic instability of flexible tubes subjected to two-phase internal flow. J. Fluids Struct. **19**(7), 943–956 (2004)

P. Montague, Experimental behavior of double-skinned, composite, circular cylindrical-shells under external-pressure. J. Mech. Eng. Sci. **20**(1), 21–34 (1978)

C.E. Murphey, C.G. Langner, Ultimate pipe strength under bending, collapse and fatigue, in *Proceedings of the 4th International Conference on Offshore Mechanics and Arctic Engineering*, vol. 1 (1985), pp. 467–477

A. Nair, C. Chauvet, A. Whooley, A. Eltaher, P. Jukes, Flow induced forces on multi-planar rigid jumper systems, in *Proceedings of the ASME 30th International Conference on Ocean, Offshore and Arctic Engineering* (2011)

C.P. Naveira, M. Lachi, R.M. Cotta, J. Padet, Hybrid formulation and solution for transient conjugated conduction-external convection. Int. J. Heat Mass Tran. **52**(1–2), 112–123 (2009)

E. Nawy, *Concrete Construction Engineering Handbook* (CRC Press, Boca Raton, 2008)

T.A. Netto, S.F. Estefen, Ultimate strength behavior of submarine pipelines under external-pressure and bending. J. Construct. Steel Res. **28**(2), 137–151 (1994)

T.A. Netto, U.S. Ferraz, S.F. Estefen, The effect of corrosion defects on the burst pressure of pipelines. J. Construct. Steel Res. **61**(8), 1185–1204 (2005)

D.J. Nicklin, J.O. Wilkes, J.F. Davidson, Two-phase flow in vertical tubes. Trans. Inst. Chem. Eng. **40**, 61–68 (1962)

R.H. Nutall, M.G. Rogers, Insulated pipe-in-pipe subsea hydrocarbon flowlines, in *Proceedings of the 17th International Conference on Offshore Mechanics and Arctic Engineering*, Lisbon (1998)

A. Nysveen, H. Kulbotten, J.K. Lervik, A.H. Børnes, M. Høyer-Hansen, J.J. Bremnes, Direct electrical heating of subsea pipelines - Technology development and operating experience. IEEE Trans. Ind. Appl. **43**(1), 118–129 (2007)

R.W. Ogden, *Non-linear Elastic Deformations* (Dover Publications, New York, NY, 1997)

A. Orell, Experimental validation of a simple model for gas-liquid slug flow in horizontal pipes. Chem. Eng. Sci. **60**(5), 1371–1381 (2005)

A. Ortega, A. Rivera, O.J. Nydal, C.M. Larsen, On the dynamic response of flexible risers caused by internal slug flow, in *Proceedings of the 31st International Conference on Ocean, Offshore and Arctic Engineering*, Rio de Janeiro, RJ (2012)

A. Ortega, A. Rivera, C.M. Larsen, Flexible riser response induced by combined slug flow and wave loads, in *Proceedings of the 32nd International Conference on Ocean, Offshore and Arctic Engineering*, Nantes (2013)

M.P. Païdoussis, Flutter of conservative systems of pipes conveying incompressible fluid. J. Mech. Eng. Sci. **17**, 19–25 (1975)

M.P. Païdoussis, *Fluid-Structure Interactions: Slender Structures and Axial Flow*, vol. 1 (Academic, San Diego, 1998)

M.P. Païdoussis, *Fluid-Structure Interactions: Slender Structures and Axial Flow*, vol. 2 (Academic, San Diego, 2001)

M.P. Païdoussis, Some unresolved issues in fluid-structure interactions. J. Fluids Struct. **20**(6), 871–890 (2005)

M.P. Païdoussis, The canonical problem of the fluid-conveying pipe and radiation of the knowledge gained to other dynamics problems across applied mechanics. J. Sound Vib. **310**(3), 462–492 (2008)

M.P. Païdoussis, N.T. Issid, Dynamic stability of pipes conveying fluid. J. Sound Vib. **33**, 267–294 (1974)

M.P. Païdoussis, G.X. Li, Pipes conveying fluid: a model dynamical problem. J. Fluids Struct. **7**(2), 137–204 (1993)

L.N. Panda, R.C. Kar, Nonlinear dynamics of a pipe conveying pulsating fluid with combination, principal parametric and internal resonances. J. Sound Vib. **309**(3–5), 375–406 (2008)

M.S. Pantazopoulos, C.W. Crossley, G. Orgill, K.F. Lambrakos, Fourier methodology for pipeline span vortex-induced vibration analysis in combined flow, in *Proceedings of the ASME 12nd International Conference on Ocean, Offshore and Arctic Engineering* (1993)

I.P. Pasqualino, B.C. Pinheiro, S.F. Estefen, Comparative structural analyses between sandwich and steel pipelines for ultra-deep water, in *International Conference on Offshore Mechanics and Arctic Engineering* (2002)

M.H. Patel, F.B. Seyed, Internal flow-induced behaviour of flexible risers. Eng. Struct. **11**(4), 266–280 (1989)

C. Paz, Estudo usando correlação numérico e experimental visando a determinação da pressão de colapso em modelos de duto Sanduíche com Núcleo de PVDF. PhD Thesis, COPPE/UFRJ. Rio de Janeiro, 2013

C. Paz, M.I. Lourenço, S. Estefen, Influência do método carretel na resistência estrutural de dutos sanduíche com núcleo de compósitos cimentícios reforçados com fibras de PVA (SHCC) (2014)

E.B. Pereira, G. Fischer, J.A.O. Barros, Direct assessment of tensile stress-crack opening behavior of strain hardening cementitious composites (SHCC). Cement Concrete Res. **42**(6), 834–846 (2012)

M.J. Pettigrew, C.E. Taylor, Two-phase flow-induced vibration: an overview. J. Press. Vessel Technol. **116**(3), 233–253 (1994)

M.J. Pettigrew, C.E. Taylor, N.J. Fisher, M. Yetisir, B.A.W. Smith, Flow-induced vibration: recent findings and open questions. Nuclear Eng. Design **185**(2–3), 249–276 (1998)

B.A. Price, T.F. Smith, Thermal response of composite building envelopes accounting for thermal-radiation. Energ. Convers. Manage. **36**(1), 23–33 (1995)

Q. Qian, L. Wang, Q. Ni, Instability of simply supported pipes conveying fluid under thermal loads. Mech. Res. Commun. **36**(3), 413–417 (2009)

L. Qinhua, X. Shilang, Experimental research on mechanical performance of hybrid fiber reinforced cementitious composites with polyvinyl alcohol short fiber and carbon textile. J. Compos. Mater. **45**(1), 5–28 (2010)

C.R. Regis, R.M. Cotta, J. Su, Improved lumped analysis of transient heat conduction in a nuclear fuel rod. Int. Commun. Heat Mass **27**(3), 357–366 (2000)

J.L. Riverin, M.J. Pettigrew, Vibration excitation forces due to two-phase flow in piping elements. J. Press. Vessel Technol. **129**(1), 7–13 (2007)

J.-L. Riverin, E. de Langre, M.J. Pettigrew, Fluctuating forces caused by internal two-phase flow on bends and tees. J. Sound Vib. **298**(4–5), 1088–1098 (2006)

M.H. Sadeghi, M.H. Karimi-Dona, Dynamic behavior of a fluid conveying pipe subjected to a moving sprung mass - an FEM-state space approach. Int. J. Press.Vessel Pip. **88**(4), 123–131 (2011)

R.C. Sarmento, G.A.S. Ribbe, L.F.A. Azevedo, Wax blockage removal by inductive heating of subsea pipelines. Heat Trans. Eng. **25**(7), 2–12 (2004)

Y.S. Seo, W.B. Jeong, S.H. Jeong, J.S. Oh, W.S. Yoo, Finite element analysis of forced vibration for a pipe conveying harmonically pulsating fluid. JSME Int. J. Ser. C-Mech. Syst. Mach. Elem. Manuf. **48**(4), 688–694 (2005)

F.B. Seyed, M.H. Patel, Mathematics of flexible risers including pressure and internal flow effects. Marine Struct. **5**(2–3), 121–150 (1992)

A. Shahba, S. Rajasekaran, Free vibration and stability of tapered Euler-Bernoulli beams made of axially functionally graded materials. Appl. Math. Model. **36**(7), 3088–3105 (2012)

A. Shahba, R. Attarnejad, M.T. Marvi, S. Hajilar, Free vibration and stability analysis of axially functionally graded tapered Timoshenko beams with classical and non-classical boundary conditions. Compos. Part B-Eng. **42**(4), 801–808 (2011)

H.-S. Shen, *Functionally Graded Materials : Nonlinear Analysis of Plates and Shells* (CRC Press, Boca Raton, 2009)

G.G. Sheng, X. Wang, Thermomechanical vibration analysis of a functionally graded shell with flowing fluid. Eur. J. Mech. A/Solids **27**(6), 1075–1087 (2008)

H. Showkati, R. Shahandeh, Experiments on the buckling behavior of ring-stiffened pipelines under hydrostatic pressure. J. Eng. Mech. **136**(4), 464–471 (2009)

M. Simsek, T. Kocaturk, S.D. Akbas, Dynamic behavior of an axially functionally graded beam under action of a moving harmonic load. Compos. Struct. **94**(8), 2358–2364 (2012)

B. Singh, P. Jukes, B. Poblete, B. Wittkower, 20 Years on lessons learned from piper alpha. The evolution of concurrent and inherently safe design. J. Loss Prevent. Proc. Ind. **23**(6), 936–953 (2010)

J.K. Sinha, A.R. Rao, R.K. Sinha, Prediction of flow-induced excitation in a pipe conveying fluid. Nuclear Eng. Design **235**(5), 627–636 (2005)

R.F. Solano, *Thermal Buckling in a Dual Pipe-in-Pipe System in Deepwaters of. M.Sc. Thesis (in Portuguese)* (Federal University of Rio de Janeiro, Rio de Janeiro, 2001)

J. Su, Flow assurance of deepwater oil and gas production—a review, in *Proceedings of 22nd International Conference on Offshore Mechanics and Arctic Engineering (OMAE'03)*, Cancun (2003)

J. Su, D.R. Cerqueira, Simulation of transient heat transfer in multilayered composite pipe, in *Proceedings of 20th International Conference on Offshore Mechanics and Arctic Engineering (OMAE'01)*, Rio de Janeiro (2001)

J. Su, R.M. Cotta, Improved lumped parameter formulation for simplified LWR thermohydraulic analysis. Ann. Nucl. Energy **28**(10), 1019–1031 (2001)

J. Su, C.R. Regis, A. Botto, Thermal Analysis of combined active heating and passive insulation of deepwater pipelines, in *Proceedings of 21st International Conference on Offshore Mechanics and Arctic Engineering (OMAE'02)*, Oslo (2002)

J. Su, D.R. Cerqueira, S.F. Estefen, Thermal analysis of sandwich pipes with active electrical heating, in *Proceedings of 22nd International Conference on Offshore Mechanics and Arctic Engineering (OMAE'03)*, Cancun (2003)

J. Su, D.R. Cerqueira, S.F. Estefen, Simulation of transient heat transfer of sandwich pipes with active electrical heating. J. Offshore Mech. Arct. Eng. **127**(4), 366–370 (2005)

Y. Taitel, Stability of severe slugging. Int. J. Multiphase Flow **12**(2), 203–217 (1986)

Z. Tao, L.H. Han, X.L. Zhao, Behaviour of concrete-filled double skin (CHS inner and CHS outer) steel tubular stub columns and beam-columns. J. Construct. Steel Res. **60**(8), 1129–1158 (2004)

J. Thomas, A. Ramaswamy, Mechanical properties of steel fiber-reinforced concrete. J. Mater. Civil Eng. **19**(5), 385–392 (2007)

G. Tough, S. Denniel, M. Al Sharif, J. Hutchison, Nile - design and qualification of reeled pipe in pipe for Deepwater, in *Proceedings of Offshore Technology Conference*, Houston (2001)

S. Trout, B. Sahota, Shell ETAP high pressure and temperature pipe-in-pipe pipeline design and fabrication, in *Proceedings of the 18th International Conference on Offshore Mechanics and Arctic Engineering*, Newfoundland (1999)

G.P.A.G. van Zijl, Modeling masonry shear-compression: Role of dilatancy highlighted. J. Eng. Mech. **130**(11), 1289–1296 (2004)

R.V. Velasco, Self-consolidating concretes reinforced with high volumetrics fracions of steel fibers: rheological, physics, mechanics and thermal properties. PhD Thesis, COPPE, Federal University of Rio de Janeiro, Rio de Janeiro, 2008

H.M. Wang, An effective approach for transient thermal analysis in a functionally graded hollow cylinder. Int. J. Heat Mass Tran. **67**, 499–505 (2013)

J.C. Wang, Y.K. Chen, *Application of ABAQUS in Civil Engineering* (Zhejiang University Press, Hangzhou, 2006)

H.M. Wang, C.B. Liu, Analytical solution of two-dimensional transient heat conduction in fiber-reinforced cylindrical composites. Int. J. Therm. Sci. **69**, 43–52 (2013)

Z.M. Wang, Z.W. Zhang, F.Q. Zhao, Stability analysis of viscoelastic curved pipes conveying fluid. Appl. Math. Mech. (English Edition) **26**(6), 807–813 (2005)

S. Wei, S.T. Mau, C. Vipulanandan, S.K. Mantrala, Performance of new sandwich tube under axial loading - analysis. J. Struct. Eng. **121**(12), 1815–1821 (1995a)

S. Wei, S.T. Mau, C. Vipulanandan, S.K. Mantrala, Performance of new sandwich tube under axial loading - experiment. J. Struct. Eng. **121**(12), 1806–1814 (1995b)

M.B. Weimann, V.C. Li, Hygral behavior of engineered cementitious composites (ECC). Int. J. Restor. Build. Monuments **9**(5), 513–534 (2003)

S. Wolfram, *The Mathematica Book*, 5th edn. (Wolfram Media/Cambridge University Press, Champaign, 2003)

H.D. Wright, T.O.S. Oduyemi, H.R. Evans, The design of double skin composite elements. J. Construct. Steel Res. **19**(2), 111–132 (1991a)

H.D. Wright, T.O.S. Oduyemi, H.R. Evans, The experimental behavior of double skin composite elements. J. Construct. Steel Res. **19**(2), 97–110 (1991b)

T. Xu, B. Lauridsen, Y. Bai, Wave-induced fatigue of multi-span pipelines. Marine Struct. **12**(2), 83–106 (1999)

M.R. Xu, S.P. Xu, H.Y. Guo, Determination of natural frequencies of fluid-conveying pipes using homotopy perturbation method. Comput. Math. Appl. **60**(3), 520–527 (2010)

Q. Xu, S. Gong, Q. Hu, Collapse analyses of sandwich pipes under external pressure considering inter-layer adhesion behaviour. Marine Struct. **50**, 72–94 (2016)

F. Yagishita, H. Kitoh, M. Sugimoto, T. Tanihira, K. Sonoda, Double-skin composite tubular columns subjected cyclic horizontal force and constant axial force, in *Proceedings of the Sixth ASCCS Conference*, Los Angeles, vol. 1 (2000), pp. 497–503

J. Yang, G. Fischer, Simulation of the tensile stress-strain behavior of strain hardening cementitious composites, in *Measuring, Monitoring and Modeling Concrete Properties* (Springer, Dordrecht, 2006), pp. 25–31

X.D. Yang, T.Z. Yang, J.D. Jin, Dynamic stability of a beam-model viscoelastic pipe for conveying pulsative fluid. Acta Mech. Solida Sin. **20**(4), 350–356 (2007)

M.K. Yeh, S. Kyriakides, Collapse of deepwater pipelines. J. Energy Res. Technol. **110**(1), 1–11 (1988)

T. Yu, J.G. Teng, Y.L. Wong, S.L. Dong, Finite element modeling of confined concrete-I: Drucker-Prager type plasticity model. Eng. Struct. **32**(3), 665–679 (2010a)

T. Yu, J.G. Teng, Y.L. Wong, S.L. Dong, Finite element modeling of confined concrete-II: plastic-damage model. Eng. Struct. **32**(3), 680–691 (2010b)

H.B. Zhai, Z.Y. Wu, Y.S. Liu, Z.F. Yue, Dynamic response of pipeline conveying fluid to random excitation. Nuclear Eng. Design **241**(8), 2744–2749 (2011)

H.B. Zhai, Z.Y. Wu, Y.S. Liu, Z.F. Yue, In-plane dynamic response analysis of curved pipe conveying fluid subjected to random excitation. Nuclear Eng. Design **256**, 214–226 (2013)

Y.L. Zhang, L.Q. Chen, Internal resonance of pipes conveying fluid in the supercritical regime. Nonlinear Dyn. **67**(2), 1505–1514 (2012)

Y.L. Zhang, L.Q. Chen, External and internal resonances of the pipe conveying fluid in the supercritical regime. J. Sound Vib. **332**(9), 2318–2337 (2013)

W. Zhang, D.X. Li, Study on first-yield collapse of submarine pipelines. China Ocean Eng. **11**(4), 449–456 (1997)

M. Zhang, J. Xu, Effect of internal bubbly flow on pipe vibrations. Sci. China Technol. Sci. **53**(2), 423–428 (2010). ISSN 1674-7321

Y.L. Zhang, D.G. Gorman, J.M. Reese, A modal and damping analysis of viscoelastic Timoshenko tubes conveying fluid. Int. J. Numer. Method Eng. **50**(2), 419–433 (2001)

J. Zhang, K.Y.C. Leung, Y.N. Cheung, Flexural performance of layered ECC-concrete composite beam. Compos. Sci. Technol. **66**(11), 1501–1512 (2006)

M.M. Zhang, J. Katz, A. Prosperetti, Enhancement of channel wall vibration due to acoustic excitation of an internal bubbly flow. J. Fluids Struct. **26**(6), 994–1017 (2010)

F.Q. Zhao, Z.M. Wang, Z.Y. Feng, H.Z. Liu, Stability analysis of Maxwell viscoelastic pipes conveying fluid with both ends simply supported. Appl. Math. Mech. (English Edition) **22**(12), 1436–1445 (2001)

X.L. Zhao, B.K. Han, R.H. Grzebieta, Plastic mechanism analysis of concrete-filled double-skin (SHS inner and SHS outer) stub columns. Thin-Walled Struct. **40**(10), 815–833 (2002)

Z. Zhong, L. Wu, W. Chen. *Mechanics of Functionally Graded Materials and Structures* (Nova Science, New York, 2012)

Index

C. An et al., *Structural and Thermal Analyses of Deepwater Pipes*,
https://doi.org/10.1007/978-3-030-53540-7

Index

C. An et al., *Structural and Thermal Analyses of Deepwater Pipes*,
https://doi.org/10.1007/978-3-030-53540-7